Are Numbers Real?

Brian Clegg

ROBINSON

ROBINSON

First published in the US in 2016 by St Martin's Press, New York

This edition published in Great Britain in 2017 by Robinson

1 3 5 7 9 10 8 6 4 2

A CIP catalogue record for this book
is available from the British Library.

ISBN: 978-1-47213-976-4

Printed and bound in Great Britain by
CPI Group (UK) Ltd, Croydon CR0 4YY

Robinson
An imprint of
Little, Brown Book Group
Carmelite House
50 Victoria Embankment
London EC4Y 0DZ

An Hachette UK Company
www.hachette.co.uk

www.littlebrown.co.uk

Brian Clegg is a prize-winning science writer with a physics degree from Cambridge and a masters in the mathematical discipline operational [illegible]. He has written over 20 science books and articles for newspapers and magazines from *The Observer* and *Wall Street Journal* to *BBC Focus* and *Playboy*. He lives in Wiltshire, England, with his wife and two children.

Other titles

A Brief History of Infinity by Brian Clegg

Ten Physicists Who Transformed Our Understanding of Reality by Brian Clegg and Rhodri Evans

A Slice of Pi by Liz Strachan

Easy as Pi by Liz Strachan

A Brief History of Mathematical Thought by Luke Heaton

A Brief Guide to the Great Equations by Robert Crease

For Gillian, Rebecca, and Chelsea

As far as the laws of mathematics refer to reality, they are not certain; and as far as they are certain, they do not refer to reality.

—Albert Einstein, *Sidelights on Relativity* (1922)

Contents

Acknowledgments

Thanks to all those at St. Martin's Press who have made this book possible, including Michael Homler and Lauren Jablonski. Thanks also to the many people who have helped me muse over the relationship of mathematics and science, notably Professor Adrian Moore of Oxford University, who was inspired to go into philosophy by the same math teacher who spurred me on to take an interest in mathematics and infinity, Neil Sheldon of the Manchester Grammar School.

Are Numbers Real?

1 Counting Sheep

Our journey in this book will explore a question that is fundamentally important to scientists—and for that matter the rest of us. Yet it's a question that most people, including scientists, rarely give a moment's thought to. Are numbers, and is the wider concept of mathematics, real?

At first glance, this seems a crazy question to devote thirty seconds to, let alone a whole book. Of course numbers are real. You only have to take a look at my bank statement. It contains a whole load of numbers, most of which seem to be negative as cash flows out of the account. And as for mathematics, we all had plenty of homework when we were at school, and that seemed real enough at the time. But here I'm using a different definition of "real." It is essential to gain a better understanding of science to discover whether numbers and mathematics form real entities, whether they have a factual existence in the universe. Would numbers exist without people to think about them, or are they just valuable human inventions, the imaginary inhabitants of a useful fantasy world?

We know that it is perfectly possible to devise mathematics that does not have any underlying link with reality. Mathematicians do this all the time. Math,* in the end, provides nothing more or less than a set of rules that are used to get from a starting point to an outcome. We can define those rules in such a way that they happen to match what we observe in the real world, or we can make them as bizarrely and wonderfully different from reality as we like. And some mathematicians delight in taking such fantasy journeys into alternative universes.

To take a simple example, the real world has three spatial dimensions (unless string theory, the attempt in physics to combine gravity and the other forces of nature that requires 9- or 10-dimensional space, has it right—see page 236)—but a mathematician is just as comfortable working with 1, 2, 4, 79, or 5,000 dimensions. Mathematicians delight in the existence of a mathematical construct called the Monster group, which is a group of ways you could rotate things if space had 196,883 dimensions. When working with the Monster, to quote Dorothy in *The Wizard of Oz,* "Toto, I've a feeling we're not in Kansas anymore."

For that matter, when mathematicians work on something as everyday as the shape of knots, they make their own definition of what a knot is that bears no resemblance to the things we use to tie up shoelaces. For reasons of practical convenience, the mathematicians set a rule that both ends of the string they

* I am conscious that this book will be read by both its original American audience and those from other countries, where mathematics is shortened to "maths." I apologize for the genuine pain I know that seeing "math" will cause those readers, but adding "(maths)" every time would be clumsy.

are knotting must be joined together, making a continuous loop. We know real-world knots aren't like that—even mathematicians (admittedly not the most worldly people) know this—but they don't care, because that's the rule that they chose to use.

Similarly we could devise a mathematical system in which $2+2=5$. It doesn't work with real-world objects, but there is no reason why it can't with a number system if we define it to work that way. Although not so extreme, there is a commonly used mathematical system where we can define $2+2$ to be 0 or 1. It's called clock arithmetic. Instead of numbers adding constantly, they progress like the numbers on a clock, resetting to 0 at a specific value. Admittedly these do have a parallel in the world. We use clock arithmetic, as the name suggests, on analog clocks. On a twelve-hour clock, for instance, $9+6=3$. Such arithmetic provides a better representation of anything cyclical than traditional counting. What this illustrates is both the arbitrariness of mathematics and how we have to be careful about definitions. The number 9 on a clock is not the same thing as the number 9 when we are counting goats, they just have some things in common, and use the same symbol.

To turn it around and consider things from the real-world viewpoint, it is possible to go through life without ever encountering much in the way of mathematics. For most of human existence, the vast majority of human beings have managed to do so. Some very basic arithmetic seems to be preprogrammed. Both dogs and babies react with surprise when, for instance, one item is put into a bowl, then another, but when they then look in the bowl, there is only one object, because the second was palmed. "$1+1=2$" seems a pretty low-level mammal programming, and is without doubt useful in calculating the odds when faced with more than one enemy to fight. Most of the rest of mathematics is a late add-on to our capabilities, but one that has proved extremely useful.

Without mathematics, hardly any of the science and technology that is essential for today's civilization would be produced. Math threads through our lives, from everyday functions like transactions in a store, to understanding the significance of the distribution of a disease or the outcome of an election. Because it *is* important that we have a feel for a discipline that is so useful in understanding the underlying structures and principles of the world around us and predicting its behavior, it's a shame that so many of us find getting into mathematics remarkably difficult, or even painful—something to be avoided if at all possible. A 2012 British article for World Math Day commented:

> We know too that many adults simply don't like maths and don't see the point of it. Many have no qualms about saying so. Being "no good at maths" carries little stigma. That tends not to be the case in other parts of the world. Negative attitudes to maths set in early in the UK—some would say between the ages of seven and nine, when many children's interest and attainment dip, in most cases never to return. They switch off and decide maths is something to be borne until the moment they can give it up—for ever. . . . The process is then cyclical, with parents (and in some cases—dare I say?—teachers) passing on their own lack of enthusiasm and confidence to the next generation.

The article suggests that this problem of having a negative attitude to mathematics is a particularly British one, but I suspect that it is one that is reflected not only in the United States, but also across many parts of the world. And this opinion is nothing new. St. Augustine of Hippo wrote back in AD 415, "The danger . . . exists that the mathematicians have made a covenant with the devil to darken the spirit and confine man in the bonds of hell." He clearly did not have much fun in his geometry lessons. (The quote is a touch misleading. Augustine was usually more supportive of mathematics—the word translated as "mathematicians" referred to astrologers—but it still reflects the feeling that many seem to have.)

And yet mathematics can be both enjoyable, when presented the right way, and immensely powerful. The fun comes from mathematical puzzles and diversions. The delight can be particularly strong with mathematics that entertains by making your head spin—like the idea that there is more than one size of infinity (see page 181).

We might not need much math to go through our basic everyday lives, and the vast majority of us get by with a touch of arithmetic and little else. But when scientists and engineers try to understand how things work and to construct products based on that understanding, mathematics has proved an invaluable tool to gain insights. Without it, it would be very difficult to understand much about the natural world, or to predict how it is going to behave. Without it, we would not have the computer this was written on or much of the other technology that supports our modern lives.

Initially mathematics was intimately tied to natural behavior. Numbers, for instance, corresponded to tangible objects. But with time it has become separated from reality. There is still applied mathematics with a tie to the real world, but pure mathematics soared in the Renaissance as mathematicians realized that they were playing an immense game where they could set their own rules, play along, and see what happened. Sometimes some of the ideas and worlds they generated would have practical uses, sometimes they wouldn't. The distinction was arbitrary as far as they were concerned (and often remains so). The great game is paradoxical in that it is both totally open and surprisingly restrictive—what mathematics covers, what rules you set, are absolutely up to you. But once those rules have been established, the game says that you must stick to them. In math you can never cheat.

When we contemplate the nature of numbers and reality, the arbitrariness that lies beneath the surface of mathematics can lead to problems for the more rigidly minded. In the British Court of Appeal, the second highest court in the UK, in 2015, three judges were set the task of deciding exactly what the number "1" meant. And their decision certainly didn't equate to something that matched the understanding most of us (or even most mathematicians) would have.

The legal case was a wrangle between two pharmaceutical companies over a chemical solution used to reduce infection in wound dressings, and bizarrely the case led to a change in the legal definition of the number "1" in the UK. The problem was that one company, ConvaTec, had a patent covering a silver-based solution of "between 1 per cent and 25 per cent of the total volume of treatment." The rival pharma company, Smith & Nephew, had therefore devised a competing product containing a 0.77 percent solution, which they believed kept them safely outside the remit of their competitor's patent.

This dispute had already been taken to trial in 2013. A lower court agreed that the "1" at the lower end of the ConvaTec range did not simply represent the numerical value "1," corresponding to a single object. Instead, they adopted an approach not uncommon among chemists, but unusual mathematically, of defining the value as the split between the ranges of two significant figures. This "significant figure" aspect meant that "1" was defined by the lower court as anything between 0.95 and 1.5—giving a conveniently asymmetric definition that left the Smith & Nephew product legal. Unhappy with this approach, the Court of Appeal judges went for a more familiar arithmetic approach of rounding to the nearest whole number, meaning that "1" now became anything between 0.5 and 1.4999. The result was to put Smith & Nephew into a difficult position. But it also demonstrates the arbitrary nature of mathematical decisions.

There is no "right" way to define something that is 1 unless you stick to exactly and only 1—and the result is that, as far as the lawyers are concerned, the range "between 1 and 25" would have to include 0.5. One of the panel, Lord Justice Christopher Clarke, made the unhelpful statement, "A linguist may regard the word 'one' as meaning 'one—no more and no less.' To those skilled in the art it may, however, in context, imply a range of values extending beyond the integer." It's not clear exactly what dark art he had in mind.

Over time, mathematicians (as opposed to lawyers) have been distinctly creative in their handling of math. The way they operate is a bit like the way that some companies allow their employees to play around with different ideas and technologies in the hope that a new product will emerge. Often nothing relevant to the commercial world will be produced. But every now and then, something wonderful and genuinely new will be brought into being. Similarly, when mathematicians played around with an idea like the square root of a negative number (see chapter 8), called an imaginary number, they were initially simply enjoying a new direction to take their mathematical game. But, as it happened, because of the rules they decided to apply to this magical class of numbers, it became a hugely useful tool for physics and engineering.

No scientist or engineer ever said prior to the introduction of imaginary numbers, "What we want is the square root of a negative number. They would really help us with this problem we've got." Similarly, no one in mathematics thought, "How can we solve this particular problem that the physicists have?" before dreaming up imaginary numbers. The mathematicians just played with the implications of their new concept and the set of rules attached to it. The applications emerged later.

Generally speaking, up to the nineteenth century, the mathematics that was needed for all of science was within the grasp of pretty well anyone who hasn't got serious problems with numbers. In my experience, as long as you can get on top of the workings of fractions (something a scary number of people never achieve), you can manage everything up to basic calculus, which sounds worse than it is. But in the nineteenth century, it is arguable that two things happened in mathematics that drove a wedge between the general public and science.

The first of these was the use of increasingly complex mathematical techniques that take a considerable amount of post-school study to get a handle on. Pick up a modern physics paper at random, for instance, and the chances are it will use at least one approach that never made it into high school math. It is no surprise that when Einstein developed his general theory of relativity he had to get help with the mathematics because he found it too difficult alone. The science he was fine with, but the mathematics had moved beyond his experience.

The second development that has made science less approachable was putting the cart before the horse. Mathematics had always been the servant of scientists, but in the twentieth century it increasingly was put in charge. Attempts to unify the forces of nature, for instance, became driven from the mathematics of symmetry, along the way becoming very difficult to explain in laypeople's terms. Another example was matrix mechanics, where a form of mathematics then largely unfamiliar to scientists, let alone the rest of us, was used to explain the behavior of quantum particles in a way that made it impossible to visualize what was happening. All that remained were the numbers and the rules to manipulate them.

There is nothing wrong with these developments per se, but they bring some unfortunate baggage along with them. If your science can't be easily described to the person in the street, then it becomes harder to justify spending taxpayer dollars on it. Physicists often point to the U.S. government funding decision in the 1990s between spending on the International Space Station (ISS) and the Superconducting Super Collider (SSC). The SSC was a massive particle accelerator, already by then under construction in Texas. Nobel Prize–winning physicist Steven Weinberg has pointed out that the SSC, larger than the Large Hadron Collider (LHC) at CERN, the European nuclear research organization, could have produced results a good ten years earlier than its rival, that would have added to our fundamental understanding of the universe. The ISS, by contrast, which won the funds, leading to the cancellation of the SSC, has given us a better understanding of space and space travel, which may be valuable in the future, but has made little contribution to scientific research.

What's more, the ISS has now cost the United States ten times the SSC budget in the process of delivering very little. But the difference is that the ISS was doing something that all the politicians could clearly visualize and that could be easily explained to the people controlling the funds. Hence its prioritization and the cancellation of the SSC. No one could explain to the politicians really what was going to be achieved by the collider, because it was too complicated to do so. The result was that the science funding went to the project with almost zero scientific benefit, rather than the one that would have retained the U.S. position as a world leader in physics.

Instead that glory went to CERN and the LHC. And yet in a way it was a hollow victory, because that same math-based difficulty was looming in terms of what the LHC was trying to achieve. We hear about "re-creating the conditions near the Big Bang" or "searching for the Higgs boson," but when it comes down to explaining to the public what these handy labels actually mean, it is very difficult. The Higgs boson will typically be described as "the particle that gives all the others their mass," which has elements that verge on the correct, but is also wrong in several different ways. However, it is hard even to describe why this wording is wrong without resorting to the language of mathematics—one that will immediately lose the audience.

What I don't want to do is sound miserable and negative about mathematics. As you will discover as we take our journey into the reality of numbers, it is a topic that is as fascinating as it is powerful. And it has been central to the development of modern society and technology. But that should not stop us asking some fundamental questions. Does modern science give too much emphasis to mathematics? Some scientists, such as physicist Max Tegmark, go as far as to suggest that the universe *is* mathematical. That numbers aren't just real, but are what make everything happen. Could it be that scientists like Tegmark sometimes confuse math and reality? Is mathematics really at the heart of the universe, or is it just a great tool for helping us understand what's going on? These are the questions we will discover answers to as we explore how math has come to rule the scientific world.

First, though, we need to go back to the very beginning of mathematical thinking. Let's take a leaf out of the mathematicians' book and play a game where we set the rules. Let's pretend that we are going to invent numbers.

2 Counting Goats

Do you ever feel the need to count your children to make sure they exist? It's unlikely. Similarly, for the early human hunter-gatherers with no concepts of large-scale projects or commerce, numbers would have had little significance. But once humans began to settle and to trade, counting and the ability to record those counts for future reference took on a new significance. A good place to start would be to undertake Albert Einstein's favorite approach to understanding. Let's do a thought experiment, in this case imagining that we will be responsible for bringing numbers into being. We don't know how numbers were invented historically for sure, but it's pretty likely to have gone something like this.

We start by counting. Strange though it may seem to us in our intensely number-oriented world, there is no need to have numbers to be able to count. This is something we will come back to in a big way when we take on set theory and infinity—where there are such things as countable and uncountable infinities, even though neither is a number. For the moment it is enough to go with the flow and see how counting in the form of a tally requires no concept of numbers.

Let's imagine I'm a prehistoric farmer, living in a society where numbers do not yet exist. I agree to lend my neighbor some of my goats. (I don't know why he might want them, not knowing much about goats or prehistoric farmers. I will leave that to your imagination.) My neighbor is my friend and I trust him, but equally I want to make sure that I get all of my goats back when he returns them. So starting with my hand spread open flat, I fold my pinkie into my palm as a goat leaves my stockade. Try it as we go along. (You will probably need to help your little finger with the other hand as it will tend to pull the next finger with it, left to its own devices. Hands make relatively poor counting technology, but they are very convenient.) Then I fold the next finger in as another goat leaves. After all the fingers of my hand are folded in, I put my thumb across them to indicate another goat. At this point the neighbor decides that he's got enough.

A few weeks later, he brings my goats back. I go through the same process as the goats are counted into the stockade. And I end up with the same result—so I know that all of the goats that went out have come back in. (If we are going to be picky, I don't know that they are the same goats, but that's not really of importance here.) At this stage I don't know *how many* goats were involved—I don't have a concept of "how many" or of numbers—but I know that he didn't cheat me. Inventing counting has proved to be a very valuable tool to deal with the world around me.

We do have some evidence that suggests this approach was taken in the form of old tally sticks. The earliest known example that may be a tally is a piece of bone with notches on it dating back around 20,000 years. Called the Ishango bone it is a fibula (calf bone) from a baboon, which has been decorated with three sets of deep, grouped scratches, adding up to 60, 48, and 60. These may well have been tallies, going beyond the simple number of fingers on a hand both in achieving a larger number and a greater permanence.

We have no contextual information to be sure that the role of the Ishango bone was as an aid to counting, we're just guessing. The scratches could simply have been a form of decoration. But in more recent prehistoric times there certainly were plenty of tally markings that were clearly being used for keeping a record. We will never know for certain when such a counting-without-numbers device came into being, but tallies have been regularly deployed for many millennia. Back in our thought experiment of inventing numbers, though, we've only just started. There is further to go than a simple tally.

Let's imagine that the next day after I lent the goats to my neighbor, my daughter asks me which goats I lent to my friend. I can't remember which they were (after a while, all goats look the same to me). So I say "It was" and run through the fingers-and-thumb-counting gesture. This is the closest I can come to identifying those goats. Some while later (and after my neighbor has borrowed goats several times more—it seems he is that kind of neighbor) a spark of genius strikes. Why should I go through the tedious process with the fingers and thumb when I want to talk about the goats that he borrowed? I just say "It was a hand of goats." He needs an extra goat? "A hand and a finger of goats." Without realizing it, I have invented numbers. Since they are based on fingers, my new numbers are literally digits.

Unfortunately this prehistoric version of me is getting on a bit, and my memory is not what it once was. For a while I've been carving notches in tally sticks to remember the goats that are involved in these strangely frequent transactions. But I have a word now for a special collection of fingers or tally marks—a hand. So why not scratch or paint a special symbol for a hand so I can see at a glance what the tally represents? After all, a whole string of notches rapidly become difficult to convert into a collection of hands. Initially it may well be that I use the tally mark that is still often used when counting in hands:

But after a while, being lazy, I would probably simplify it to a related squiggle, providing a downstroke to represent the fingers and a cross stroke for the thumb.

With the impact of writing and laziness, numbers have started to take on a life and a symbology of their own. And so powerfully simple is this kind of number system that you, never having used or seen it before, should be able to pretty quickly tell me what this number is:

Yes, it is what we would now call twelve. But we're getting ahead of ourselves. From our prehistoric viewpoint it is simply "hand hand finger finger." Or possibly "finger finger hands, finger finger," because I can count off the number of hands on the fingers of my left hand while counting spare fingers on the right if I want to. Woo-hoo! Am I a mathematician now? Not yet. But I am an arithmetician, if there is such a word (there appears to be as my spell-checker hasn't complained). Arguably what I am doing is probably too simplistic to be regarded as mathematics. But before we get into more sophistication, let's explore these new abilities a little further.

What I have invented is a kind of token system where my numbers stand in for real objects—they are visual representations of something physical in the world. To be more precise, in this example, they are finger positions that stand in for goats. It might seem obvious to us now that once we have these visual tokens they will work as well for, say, bags of grain as they do for goats, but that leap of abstraction is one that we know proved a struggle for early would-be mathematicians. The fact is that the universality of a number—the distancing from actual objects to independent tokens that allow us now to apply "4" as easily to cars as sausages—is not inherently obvious.

When the ancient people of the city-state of Uruk began to write, they frequently concerned themselves with accounting, just as happened in our goat thought experiment. Uruk was one of the first cities; its remains in modern Iraq. Dating back nearly as far as 4000 BC, Uruk was at the heart of the Sumerian civilization, lasting for more than 2,000 years. But the inhabitants of Uruk didn't have a single number representation that worked for everything. They had some generalization, but as far as they were concerned, some types of objects were so different that they needed their own special numbers to represent them. So, for instance, one number system was used for humans, living animals, and dried fish (don't ask), while another was employed for grain products, cheese, and fresh fish.

Nevertheless, once numbers were in play, it was inevitable that someone somewhere would eventually make the leap from numbers as, say, a measure of goats alone, to applying those same numbers to *any* other set of objects. Numbers had become universal. I am laboring this point, because it is perhaps at the heart of the issue that lies behind the question "Are numbers real?"—which is to ask why is it, if it turns out that numbers *aren't* real, they are so good at representing reality? The American mathematician Richard Hamming said:

> I have tried, with little success, to get some of my friends to understand my amazement that the abstraction of integers for counting is both possible and useful. Is it not remarkable that 6 sheep plus 7 sheep make 13 sheep; that 6 stones plus 7 stones make 13 stones? Is it not a miracle that the universe is so constructed that such a simple abstraction as a number is possible? To me this is one of the strongest examples of the unreasonable effectiveness of mathematics. Indeed, I find it both strange and unexplainable.

After a while, back in our thought experiment, I might have based a variant of my finger-counting system on physical tokens. They could be counting stones, the "calculi" that gave their name to calculations and calculus, or counters on an abacus—or, for that matter, the very specialized counting stones that we still use today and call coins. In fact I almost certainly would need to produce some sort of physical tokens fairly soon after developing numbers. My written symbols are fine for bookkeeping—to let me know just how much I've lent to my neighbor, for instance. And that's okay because he is my friend and we trust each other. But if I weren't a trustworthy person, there's nothing to stop me adding a couple of extra notches to my tally.

Now, when he brings my goats back and we count a hand of goats back in, I would pretend to be deeply wounded. "You've only given me a hand of goats," I would say. "Where are the finger finger of goats that I also lent you?" And I would show him my modified tally stick with an innocent but hurt expression on my face. Because he would have no way of backing up my tally, he is unlikely to have any recourse, other than giving me a pair of goats I never owned, or hitting me.

In reality by now, incidentally, I would probably have got fed up of saying, for instance, "finger finger" for what we would call two. So, being inventive, I would have come up with words for the intermediate values between finger and hand. As someone who works with words, and wanting to keep things short and sweet, I might end up counting: "Fin, ger, nuc, cul, hand!" So I would actually have asked "Where are the ger goats that I also lent you?" With the appropriate hurt expression.

However I represented those extra goats, I have unconsciously and without effort undertaken a new arithmetic operation. The total goats according to my fraudulent tally arc hand ger (in fact, I might run the words together to make handger for what we would call seven). And my neighbor brought me back hand goats, leaving ger goats missing. If I have hand of handger there's ger left to come—or to put it another way, handger take away hand equals ger. We're doing sums.

So, if I weren't trustworthy I could use my tally and my new-found skills in arithmetic as a con artist. Luckily my neighbor is clever enough to realize the risk, and so, instead of relying on my easily modified tally, where a goat is just represented by a notch on a stick, he provides me with a set of actual tokens. These are physical objects that he has made and will recognize, which I would find difficult to reproduce—one for each goat. Then, as he brings my goats back, I give him a token in exchange for each goat. Once hand goats are returned, I've given him all his tokens, so I can't cheat him. But here's the thing. I'm getting a bit fed up of being without my goats all the time. So we make a new arrangement. I get to keep one of his tokens as recompense for all this borrowing and some time in the future, I can exchange that for something else from him—a bag of flour, perhaps.

A combination of very basic arithmetic and not really trusting each other has produced, without any real effort, the existence of money and the ability to pay for services. They are scarily powerful things, these numbers. However, the money aspect has got us away from the purity of arithmetic, so let's just go back to considering what "hand" means. To start with, I might only use it to indicate a particular sized group of goats. But it wouldn't be long before I discovered universality, realizing that I could just as easily refer to hand apples or hand people or hand spears—as a number, hand is wonderfully versatile because it can tell us how many anythings we are dealing with.

For now, and frankly for a long time to come, that is all that hand would be or that it would need to be. It is incredibly useful for stocktaking and loans and buying and selling. It's useful to know how many are joining us for dinner so we can plan our menu, or how many nights have to pass before the days start to lengthen again. Which is why it's quite something that one of the earliest examples we have of written numbers, moving on from a simple tally, are the surprisingly advanced base-60 numbers of the Babylonians.

That "base 60" part refers the point at which the numbers move on to the next level. Today we use a base-10 system, in all likelihood originating from the fingers and thumbs of both hands. (The hand system we have developed in this chapter is technically base 5.) What was impressive about the Babylonian system, which was written in cuneiform, a script based on characters made with the end of a stylus pushed into clay tablets, was that it was a very early positional system—one where the position of a number in a row of numbers shows whether "1" is just "1" or "60" or "3600"—a system that came into use more than 2,000 years before such systems were common.

Number systems with base 5 or 10 or 20 (think fingers and toes) make a fair amount of sense, but the base 60 at first seems odd. However it turns out to be a very flexible number. It is divisible by 1, 2, 3, 4, 5, 6, 10, 12, 15, 20, and 30, which is handy when you are dividing things up. And before you dismiss base 60 entirely, bear in mind that we still use that same system for seconds in a minute, minutes in an hour, and in the way we represent angles. The Babylonians dedicated vast quantities of their clay tablets that they used as a cuneiform writing material to numbers. Many of these were for accounting purposes, and for controlling trade, but others cropped up in their work on the heavens, as the Babylonians made detailed studies of the skies, primarily for astrological reasons.

Back with my prehistoric thought experiment, the numbers we have invented stand in for real objects and are meaningless without a real object to represent. They are more like an adjective than a noun. I can't give you handger. I can only give you handger goats or handger baskets. I can't show you handger either. You might think I can show you the number when I draw the appropriate symbols, but that isn't handger any more than a sketch of a goat is a goat. And plenty of people tend to think of numbers in this adjectival fashion and in no other to this day. Many of these will be the people who struggled with math at school. Because this kind of direct representation of physical objects is just a starting point for numbers.

It is the same correspondence of numbers to objects that explains the clumsiness of many early numbering systems, which fell far behind the older Babylonian approach. The Ancient Greeks, for instance, used their standard letters of the alphabet to represent numbers, though they had to bring back a few old extinct numbers like the digamma (looking like our capital F) to have enough to go around. This confusing approach, where it was necessary to distinguish between letters and numbers by context, reflected an early separation of the two activities of writing and bookkeeping. The concept seems to be derived from the Phoenicians, and also turned up as a result in the early Hebrew representation of numbers.

For most of us, Roman numerals are more familiar, and these used a mix of straightforward tally marks and a Greek-like letter system. In fact, they even had a strong correspondence with our imagined hand-based number system that is just tally marks with I for finger and V for hand. Where the Greeks had, for instance, a separate letter for each multiple of 10 and 100, the Roman system simply repeated a letter, but with an interesting twist that gave some significance to position because the letters had to appear in decreasing order of value. If a smaller number appeared before a larger one, this was a sign that it was a modifier to be taken away.

So, for instance, think of a clockface with Roman numerals. What would you expect to see at the "4" position? If you know your Roman numerals you would probably say IV, where V indicates 5, and the "I" before it means "1 less than," getting us to 4. As it happens in this particular instance you would be wrong, as there is a strange historical convention for no good reason that on watch and clockfaces, 4 is represented by the more basic IIII (even though 9 is still IX)—but generally speaking your interpretation would be correct.

To our eyes, the systems used by both the Romans and the Greeks were extremely clumsy. They had taken a huge backward step from the Babylonians. Okay, base 60 was a little tricky to get your head around. This reflects the nature of our short-term memory. It can only cope with around 7 to 8 items at a time. This is why phone numbers, which are usually longer than 7 digits, are traditionally broken up into blocks. So number systems based on 5, or at a stretch 10, are easier to cope with than those based on 60. But there was so much about the older system that was better than what came later.

The Romans certainly had very little going for them. Their numbers were unnecessarily bulky. Compare their version of the year number 1999, which comes out as MCMXCIX, with ours. (Though just occasionally they win, such as the more compact MM for 2000.) You might wonder why we still occasionally use Roman numerals. I suspect it was because until the twentieth century there was an unjustified awe for classical culture—the same reason that the classical architectural style, considered ugly for many centuries, received a renaissance. About the only thing that Roman numerals have going for them is that they have fewer curves than our modern Arabic/Indian numerals, so are easier to carve in stone.

Most strikingly, the real problem with Roman numerals is that it makes basic arithmetic much harder than it needs to be. It isn't practical to have a simple rule (and mathematicians love a simple rule) to add XXIII to XLIV for instance, in the way that we can easily teach children how to add 23 to 45, because we have a position-based system where the column the number is in indicates how many powers of 10 it represents. This transforms the mechanics of arithmetic. (More on this in chapter 6.) It is perhaps no coincidence that the Romans who had such a poor number system never really got the hang of science.

In classical times, just as with the original tallies of goats and the like, the Greek and Roman number systems weren't designed to be mathematical tools and were not manipulated by mathematicians the way our current numbers are. Although the Greeks certainly had both a mathematical and scientific tradition, which we'll explore soon, they treated numbers very differently to the way we now do. For the moment, let's stick with the kind of basic working that was available to our imaginary prehistoric goat owner with a tally system.

You might think that negative numbers would be a step too far for such early arithmeticians. And in the modern sense, they were. You can't have –2 goats. I can't show you –2 anything. So –2 can't provide a direct representation of objects in the physical world. However, it is a concept that would be valuable to bookkeepers, even though that concept wouldn't initially be represented as a negative number. When I, as a prehistoric farmer, lent my neighbor a hand of goats, my flock was, until he returned them, missing a hand of goats. And though what was being recorded was the absence of a standard (positive) hand of goats, the notches that I made on my tally stick, or with the fingers I curled, effectively represented negative goats. When the goats were returned, each of the goat-shaped holes was filled by a returning positive goat until all the holes were filled. Just as the first use of tallies enabled us to count without numbers, so tally-based arithmetic allowed us to have negative values without recognizing negative numbers.

Even in their absence, though, the goats remained at the center of things. Whether a hand is negative or positive, it still refers to a collection of physical objects that is located somewhere in the world. But for mathematics to take off and soar it had to be detached from its linkage to sets of worldly items.

The realization came to different civilizations at different times, but in the Western tradition, the ones who first saw the light were the Ancient Greeks. It's quite popular these days to be a little dismissive of Greek science, because they didn't take what we would now consider a scientific approach and they got many things wrong. And it is certainly true that their mathematics had significant limitations. But if they had done nothing else, we would still have to thank the Ancient Greeks for making arguably the biggest single step in the development of mathematics—making it explicitly clear that numbers did not have to correspond to a specific, real-world object. For one Ancient Greek cult, whole numbers took on a whole new significance devoid of a link to counting things.

3 All Is Number

The title of this chapter is a translation of the motto that legend tells us was carved over the door lintel of the school of the Pythagoreans. For most of us, the only real contact we are likely to have had with this decidedly strange ancient Greek cult is what's now known as Pythagoras's theorem. The basic concept that the squared length of the longest side of a right-angled triangle is equal to the square of the length of the other two sides comfortably predates Pythagoras, but what is probably fairly attributed to Pythagoras, or at least one of his followers, is the proof of this idea.

Geometry, which we'll cover in more detail in chapter 5, seems to have been one of the earliest mathematical excursions after whole-number counting. We don't know for certain where or how it originated. The Greeks mostly thought it came from the Egyptians, though there are variants of the concepts of simple geometric relationships, like Pythagoras's theorem, originating earlier from Babylon, Mesopotamia, and the East. Perhaps a useful indication of why geometry arose fairly early is that an old Egyptian term for those who practiced geometry was "rope stretchers," implying that the discipline belonged to surveyors involved on building sites or in dividing up lands. (Our word geometry comes from Greek words for "Earth" and "measuring.") But the Pythagoreans did not practice such hands-dirty practical mathematics.

These are the early days of Greek mathematics—Pythagoras was born around 570 BC and is thought to have traveled to Egypt, where he may have picked up some of the ideas that later became part of the belief system of his group, the main members of which were known as the mathematikoi. As we have seen in our thought experiment, the origin of numbers was likely to have been intimately linked to the physical world. What the Pythagoreans managed to do was both elevate this concept to a fundamental, and yet at the same time to establish a distance between mathematics and simple numbers that made it possible to see math as something that could be performed in isolation from the world, without a direct physical application.

The Pythagoreans arguably forged the first link in the chain that leads to modern mathematics and science, by going beyond counting to understand and make use of the power of pattern. In a sense all human beings—indeed all living things that respond to their environment—make use of patterns. Life would be far too complicated if we had to learn how to do everything afresh each time. Instead we make use of patterns both to recognize what is around us and to shape the way that we respond to what we detect. Imagine, for instance, I programmed a robot to be able to open the window in my bedroom. As it happens it is in the form of a pair of doors opening onto a Juliet balcony, so

I would have to program the device to turn a key in a lock at a certain position on the wall, push down on a door handle located at a different position, and pull open the door with the right amount of force to open it without causing damage.

If I now moved the robot to the living room, which happens to have the same kind of window, but in a different location, my robot's precise programming would mean that it would miss its target and fail. Of course things would be far worse if, for instance, the room had a sash window. We learn the patterns of things like windows and doors and then are able to cope with them accordingly, without having to learn how to deal with every single window separately.

The same kind of process is at work when we try to understand the universe. All of science is based on making use of patterns to simplify the process of comprehension. If we had to have a separate study of every single atom to understand what was going on, we would never get anywhere. But if we can build up a pattern of what atoms in general are like, and apply that pattern across the board, we can make progress.

The Ancient Greeks had not yet come up with the concept of atoms (taken from the Greek *atomos* meaning "uncuttable") when the school of Pythagoras flourished, but they still had a powerful feel for the importance of pattern, whether it was the shape patterns of geometry, the harmonic patterns of music, or some of the patterns of numbers. As we will see, they also took those patterns too far, something humans are always prone to do. We are very good at seeing a pattern where one does not exist. It could be a visual pattern—when we see a bogeyman in what is only a collection of shadows—or a statistical pattern, when we expect there to be a cause for a cluster of events, even when randomness inherently produces such clusters with no cause.

Bear in mind that the Pythagoreans were not just mathematicians, but were also members of a cult. They had some distinctly odd beliefs, for instance a strong aversion to eating beans. It has been suggested that this was for symbolic magical reasons, because the beans looked like human organs, making eating them close to cannibalism and unsuitable for the vegetarian cult.

As far as the Pythagoreans were concerned, many of the most important patterns revolved around whole numbers. They were of the opinion that everything in the universe was structured on numbers—that numbers were not merely a human creation, but provided the underlying architecture of reality. The numbers were endowed with characteristics that made them seem like living things in their own right. The number 1, for instance, was linked to the mind and its singular nature. Two represented opinion, something that was deemed appropriate sharing—opinion. Three was linked to wholeness. This was because anything whole required a beginning, a middle, and an end, and physical objects needed three dimensions to define their existence.

And so it went on. Because odd numbers were considered to be male and even numbers to be female, 5, for instance, was an indicator of marriage (that "marriage" may be a Victorian euphemism in this context), because it joined the first true odd number, 3 (1 was considered too unique to be in the list) with the first even number, 2. This whole structure of number associations came to a climax with 10, which represented perfection. Not only was 10 the sum of the first four numbers, but, when arranged as a series of dots, it formed a perfect equilateral triangle.

Equally, the Pythagoreans were aware of the appearance of numerical patterns in music, not just in rhythm, but in the relative length of strings or pipes that produced harmonies that were pleasing to the ear. Doubling the length of a string, for instance, produced the similar-sounding octave note, while they also established the ratios required to produce other notes that sounded harmonically pleasing together. This study of musical mathematics may seem relatively trivial, but it has been suggested that it is a real milestone in the development of science, as this may well be the first example of numbers being used to derive a kind of scientific law. The study of nature, which to date had been entirely qualitative was taking on a quantitative side.

As an excellent example of the dangers of putting too much faith in the linkage between mathematics and reality, then making deductions from it (at least according to the not entirely unbiased Aristotle, who could be scathing about the older school), one Pythagorean named Philolaus used his school's obsession with the perfection of the number 10 to argue that there had to be an extra planet. This was because when he added up the Sun, the Moon, Earth, the known planets, and the sphere of the stars, making 2 + 1 + 5 + 1, he got a total of 9. As the universe had to be balanced perfection—to match the pattern that the Pythagoreans expected—Philolaus is said to have argued that the real value of the bodies must be 10 because of the number's importance, hence the unknown planet.

It would be easy now to make the association of this planet with the yet-to-be-discovered Uranus (ignoring Neptune), but Philolaus had a more dramatic, if more physically unlikely, idea that there was a "counter Earth" balancing out our home world. The more familiar later Greek cosmology that made the universe a bit like the solar system, but with the Earth at the center, had yet to be dreamed up. At the time, the prevailing model for the structure of the universe was that there was a central fire that gave light to the rest of it—the counter Earth was assumed to be on the far side of the fire, so was never seen.

Although a planet like this would return in fiction (not to be confused with the imaginary planet Vulcan, once thought to exist inside the orbit of Mercury) there has never been any astronomical evidence for its existence—it was purely a case of basing a model of how the universe works on a form of mathematics, then making deductions from it. Although this is also the way much modern physics is undertaken, we use better math—and after making an assertion, we then attempt to verify the prediction through observation or experiment, something that wasn't possible at the time for the counter Earth theory.

Nonetheless, even with a few hiccups along the way, there was just too much going for patterns in getting a feel for the nature of the universe for the approach not to be valued. Some patterns are so obvious that you can't avoid them. We are all very familiar with the repeating patterns of day and night, for instance. Some of our time patterns—the division of the day into hours or the collection of seven days into a week (based on the planets, including the Sun and the Moon), for instance—are artificial and meaningless outside human usage, but others like the day and the year are natural occurrences, real patterns of nature.

We see patterns in the way that the Sun travels across the sky—always in the same direction, at roughly the same speed—and there are other patterns that play out over a longer period in the way that the Sun's arc gets higher or lower, or in the repeated movements of the celestial lights of planets and stars. And then there are the related patterns of seasons and the longer-term patterns of life itself. It's hardly surprising that the Greeks adopted a pattern-based approach in explaining their world.

A rather less dramatic, if more romantic number-based concept that also seems to have originated with the Pythagoreans was the linkage of the so-called amicable numbers, 220 and 284. Such is the amorous appeal of this pair of numbers (when seen through Pythagorean eyes, where whole numbers were so significant) that you can even buy jewelry consisting of a metal heart broken into two, one piece with the number 220 inscribed on it and the other bearing the value 284.

The reasoning behind this is a venture into mathematical nerdiness of the highest degree. If you list the factors of 220 (the whole numbers that divide into it), you get 1, 2, 4, 5, 10, 11, 20, 22, 44, 55, and 110. Add those together and the result is 284. By comparison, the significantly fewer factors of 284 are 1, 2, 4, 71, and 142. Add those together and you get 220. It's a reciprocal linkage that the whole-number obsessed Pythagoreans could not resist.

To us this linkage is just a bit of fun—an opportunity to show off when someone asks "Why 220?" or to have a little personal secret. But clearly, with the Pythagorean view that the whole universe was constructed on numbers, these were numbers that were considered genuinely important, joining the pantheon of significant numbers.* Once you allow mathematics, and specifically whole numbers, to give you your entire viewpoint on the universe, you risk losing perspective.

This became clear in the form of a serpent in this mathematical Eden where whole numbers controlled everything. An unwanted guest that legend tells us resulted in murder.

It doesn't take a huge amount of imagination to believe that

* Some mathematicians argue that all numbers are significant. Otherwise, if you imagine the smallest number that isn't significant, then it becomes significant because of being in this position.

the cultist Pythagoreans would be prepared to kill to preserve intact their obsession with the significance of numbers, and, according to legend, this is exactly what they did. The victim was one of their number named Hippasus, who dared to reveal to the world the dirty mathematical secret that they had discovered. It's a bit like a major religion discovering that one of their sacred texts revealed that their religion was based on fiction. You can imagine the attempts that would be made to cover this up. So it's easy to understand the discomfort that this discovery caused the cult.

It all began with that apparently innocent theorem attributed to Pythagoras, specifically when it was applied to the diagonal of a square that had unit sides. The sides don't have to be any particular unit long in practice. They could be 1 centimeter or 1 mile. But the idea that they are of length 1 keeps things simple. Of course they can be any length you like and we can define a new unit based on that length, so that the sides of the square are now of unit length.

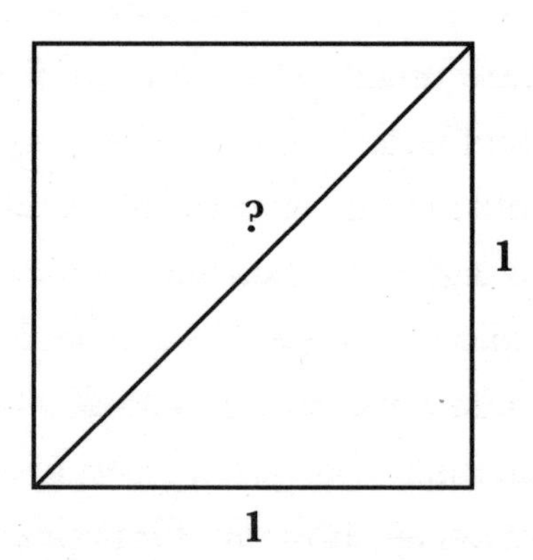

We draw a diagonal line across the square. So far, so innocent. According to the Pythagoreans' favorite theorem, it's easy to calculate the length of that diagonal, because it forms the longest side of a right-angled triangle (in fact there are two identical right-angled triangles to choose between). This means that, thanks to the power of math, we know straight away that we can find the square of the length of that diagonal by adding together the square of the other two sides. Because we handily chose the sides to be 1 unit long, this means the square of the diagonal is $1^2 + 1^2$ or $1 + 1$—better known as 2. And equally easily, by definition we would now call the length of the diagonal, the value we are trying to find $\sqrt{2}$—the square root of 2. The number that when multiplied by itself makes 2.

This all seems highly inoffensive, but things started to fall apart when the Pythagoreans tried to calculate exactly what $\sqrt{2}$ was. Bear in mind that their whole universe was constructed on the concept of numbers. Whole numbers. So everything, including $\sqrt{2}$ had to be calculable from whole numbers. Now $\sqrt{2}$ clearly isn't 1, as 1 multiplied by itself is just 1, not 2. And similarly $\sqrt{2}$ can't be 2 because 2^2 is 4. That wasn't a problem, because the Pythagoreans were also aware that there could be something in between 1 and 2 that was a ratio of two whole numbers—what we would call a fraction.

If the Greeks had had a true concept of a fraction, then the Pythagoreans would already have made an even bigger abstraction than recognizing the existence of whole numbers without specific physical objects to count. When we use the positive integers, the whole numbers, they can always be returned to the consideration of actual objects. If I say I am giving you three goats and you take one away, we are left with exactly two goats. But even making the step to a simple fraction such as a half (1/2), we are no longer quite representing reality with the same degree of perfection. Yes 2 goats are exactly 1/2 of 4 goats. But 1/2 a cake, say, can only ever be an approximation in the real world—it is about dividing something in two, where the two segments are similar, but are impossible to make identical.

Fractions inevitably come to mind when dividing up an object like a cake—but they also start to become important when we deal with spatial measurement. When first we start to divide up fields or measure out the size of a stone block to build a structure, we can make use of the equivalent of integers in some standard unit of measurement, usually initially based on an aspect of the human body like a thumb's length or a foot or a cubit (the distance from the elbow to the tip of the middle finger) or a pace (*passum*), which can be multiplied up to a thousand paces (*mille passus*)—which is a mile. But unlike goats, it is quite easy when measuring the size of a piece of stone, for instance, to use up, say, seven thumbs and then have part of a thumb still to measure. Something between nothing and a whole thumb. Something that needs a concept like a fraction.

However, the Greeks didn't make the full extra leap of abstraction in considering fractions as separate numbers because they didn't use them as we do. To begin with, they didn't use the same symbolic approach. To represent, for instance, 1/4, they would simply show the number 4, itself confusingly represented by the letter delta, with a mark like an acute accent over it. To be even more confusing, for obscure historical reasons, while the second letter of the alphabet, beta was sensibly 2, a beta with a dash over it was the symbol for 2/3—an oddity that may well derive from Egyptian practice, where 2/3 seems to have been one of the few fractions they recognized with something other than 1 on the top, and so was given a special symbol.

Similarly, a special, nonalphabetic lightning-like symbol was reserved by the Greeks for 1/2 (this wasn't entirely standard—there were several alternatives). Because of this, performing arithmetic with fractions was a nightmare. The usual approach was to buy a book of tables that would tell you that, for instance 1/2 + 1/6 was 4/6 (or 2/3) because there was no mechanism like we have to multiply top and bottom until we get numbers we can work with. Greek fractions had no explicit top and bottom, even though they were thought of as ratios of whole numbers.

When we consider the Ancient Greek approach, though, we shouldn't consider it to be entirely simplistic. The Greeks were aware of some quite intricate implications of working with fractions. A philosopher born a few decades after Pythagoras, Zeno, even made something of a name for himself by producing a paradox that depended on an odd behavior of some fractions when added together. Zeno belonged to the Eleatic school (based in Elea, the ancient city just outside the current Castellammare di Velia in Italy). The school considered change and movement to be illusory. This was all very well, but experience seemed to show something quite different, that change was happening all the time. To defend his colleagues' viewpoint, Zeno came up with a whole list of paradoxes to demonstrate that our ideas of change and movement were flawed. And this, as far as the Eleatics were concerned, meant that the experience on which most people based their ideas of change was also flawed.

Probably the most famous of Zeno's paradoxes concerns Achilles and the tortoise. These two are to have a race—Achilles, the superstar of his day, running against the slow, lumbering tortoise. It was not exactly a fair pairing, yet Zeno claimed that, given a little consideration on the part of Achilles, the hero would be unable to catch up with the plodding animal. All that was required was that Achilles gave the tortoise a lead—let it start a minute or two before he did. He was a hero, after all, so this wasn't too much to ask of him. And now, Zeno argued,

Achilles would never be able to overtake his opponent, even though he ran much faster.

To illustrate Zeno's argument, let's assume that Achilles goes at twice the speed of the tortoise. In practice he would be much faster than that, but it keeps the math simple and the same argument will apply however fast Achilles runs. Our hero waits until the tortoise has traveled 1 yard. Then he sets off. Very soon he will have covered that yard, but of course in the time he took to get to that point, the tortoise will have moved on. It will have covered 1/2 a yard. Even quicker, Achilles will cross that 1/2 a yard. Only to find the tortoise is now 1/4 of a yard ahead. He still hasn't caught it. And so it goes on. Forever. Achilles will never catch the tortoise, because whenever he gets to the point that the animal had reached, it has always moved on a little further.

It's not clear if Zeno really was unaware of what was happening here, but Ancient Greek mathematics was perfectly able to explain this oddity—in fact the explanation was made easier because of a peculiarity (from our viewpoint) of the way that they considered fractions, and because they had a very visual approach to mathematics, as was clear from their relentless enthusiasm for geometry. This was because the Greeks were so obsessed with whole numbers. So by "2" (represented by beta), they really meant "a collection of two units." It is too clumsy to say, but it is a subtly different concept from the one we have now.

This distinction was even clearer when it came to fractions, where whole numbers managed to keep their significance. Rather than thinking of 1/2 or 1/3, the Greeks would consider "the second part" or "the third part." Where we would think of 1/2 as a single object (1) split into 2, the Greeks were thinking of two whole objects (the parts) that when *put together* made up 1. So they totally avoided the abstraction issues that accompany

1/2. For instance, the idea of it being an approximation when dealing with a cake didn't occur because instead they were thinking of two whole pieces of cake that made up between them a single, bigger whole. A good way of thinking of their approach to the set of distances in Achilles and the tortoise, which we would represent in our modernist algebraic representation as

$$1 + 1/2 + 1/4 + 1/8 + 1/16 \ldots$$

is as a box that is half filled by a unit-sized stone. We then add in a stone that it would take two of to be the size of the unit stone—a stone, which is the second part in size. Then a stone we would need four of to be the same as the unit stone—the fourth part. And so on. The collection of stones would get closer and closer to filling the box, but they would never quite make it. This is the legacy of the Pythagoreans' obsession with whole numbers. The Greeks weren't thinking of fractions as we do, but rather of whole objects that would fit into another object two or three or four times.

We would now say that the series below tends to 2:

$$1 + 1/2 + 1/4 + 1/8 + 1/16 \ldots$$

Each item added to the sequence brings it closer and closer to 2, though it never quite makes it. In principle, if you had the whole infinite set of the sequence in the series then it would add up to 2. But it would never add up to more than that. We now say that the total tends to 2 as the number of items in the series tends to infinity. So in the real world, the tortoise would only cover 2 yards before Achilles powered past it and the paradox collapses. Arguably, the visual Greek approach to the sum of the series is easier to accept than our modern series, though, because it is clear that as we add the narrower and narrower stones, we will never quite fill the box. It isn't so obvious when looking at that sequence

$$1 + 1/2 + 1/4 + 1/8 + 1/16 \ldots$$

that the outcome is going to be finite. In fact the remarkably similar series

$$1 + 1/2 + 1/3 + 1/4 + 1/5 \ldots$$

doesn't "converge" as the mathematicians call it and would tend to an infinite sum as the number of items tends to infinity.

Meanwhile, back with the Pythagoreans, who were trying to decide what ratio of whole numbers made up the square root of 2. If you play around with fractions, the appropriate value appears to be somewhere in the region of 707/500, but it isn't quite that. Strictly speaking, that would be 707 lots of 1/500 to the Greeks, though to make matters even more confusing it would have been represented as something like 1, 1/5, 1/5, 1/100, 1/500, 1/500. After a certain amount of head scratching and throwing a spot of logic at the problem, the Pythagoreans were able to prove that √2 was not the ratio of *any* two whole numbers. It couldn't be done. This new value didn't fit into their worldview. Their whole number system for the universe was shattered.

It sounds quite a challenge to prove that there was no appropriate ratio of whole numbers to produce √2. At first sight it seems to require the Ancient Greeks to test out every one of an infinite set of different fractions to make sure there wasn't a match—clearly something that was impossible to do. However there was nothing the Greeks liked more than a spot of logic, and this was a proof that could be made using only basic knowledge of the behavior of odd and even numbers and an elementary working of logic. The argument uses a popular classical approach of assuming that something is true and showing that the result of that assumption is an impossible outcome—so it can't be true. And the proof goes something like the following.

Assuming that there exists a ratio of whole numbers that is the length of the diagonal—let's say that it is α/β to use handy Greek letters—then we've got two whole numbers, α and β, and $\alpha/\beta = \sqrt{2}$. To keep things simple, we will make α and β the smallest whole numbers that produce this ratio. So they would be like 2/3 rather than 4/6 or 200/300.

We'll do the old trick of multiplying both sides by β to get rid of the messy division, giving

$$\alpha = \beta \times \sqrt{2}$$

And let's get rid of that square root by multiplying each side by itself:

$$\alpha^2 = \beta^2 \times 2$$

So far we've used basic high school math (the Greeks would have done it a little differently, but the outcome would be the same). The Greeks knew that anything multiplied by 2 is even. So the right-hand side of the equation is even . . . which means α^2 is even too. Which means α itself is even—because multiplying an odd number by itself would produce an odd value.

Anything even can be divided by 2. So α^2 can be divided by 4. Which means $\beta^2 \times 2$ can be divided by 4 as well. So β^2 can be divided by 2. Which means β^2 is even . . . and so β is even and is itself divisible by 2.

Finally we get to the point. We have shown that both α and β can be divided by 2. Which means the values used in α/β can't be the smallest whole numbers to make this ratio as was originally specified. So we have established that the smallest whole numbers *aren't* the smallest whole numbers. The original statement doesn't make logical sense. Which means that we can be sure that there is no ratio of whole numbers that results in $\sqrt{2}$.

Now this doesn't bother us. We just shrug our shoulders and stick with $\sqrt{2}$, or write out as many decimal places of the diagonal's length as we require, such as 1.414213452 . . . but these were not options that were available to the whole-number obsessed Pythagoreans. We now call a number like this irrational because it can't be made from the ratio of two whole numbers, but for the Pythagoreans it seemed to literally be an assault on the foundations of rationality. At least this would have been the case if they really associated numbers with geometrical shapes of this kind. The story of the revenge on poor Hippasus, who was said to have been taken out in a boat and drowned for giving away the secret of irrational numbers, is entertaining, but in reality, geometry back then was seen as a totally different exercise to the handling of numbers, and it is quite possible that the Pythagoreans simply saw this as a difference between shapes and number. Geometry was a visual concept as far as they were concerned, not truly a numerical one.

We now very happily make use of irrationals, and in fact any "real numbers" as we now refer to the non-integers, whether irrational like the square root of 2 or rational but represented as a decimal sequence, such as 0.666666 . . . for 2/3. In a world where computers and phones and digital displays have largely replaced manual calculations, these real numbers proved the easiest way to go. They are usually rounded to fit the capabilities of the device, so 0.666666 . . . (where the ". . ." means that the number goes on forever) becomes 0.666667, rounded up in this case on the final digit.

Generally speaking it isn't clear whether such real numbers are, indeed *real,* in the sense of being directly related to the nature of the universe rather than being a construct of mathematics. The ratio of the circumference to the diameter of a mathematical circle is the irrational number pi, but in the real world, such perfection is never achieved or achievable. Even if, somehow, a circle could be drawn perfectly, there is a granularity in matter, reflecting the way that it is made up of atoms. However the circle is physically constructed, whether it is a circular piece of metal or a shape printed on a page, this lack of perfect continuity in the real world means that we can never achieve such an idealized match between what we construct and the predictions of mathematics.

We can, of course, do away with matter and just consider space itself. Current physics suggest that even here there may be granularity, that space itself may not be totally continuous, but is probably quantized at the level of the Planck length, an immeasurably tiny size around 10^{-35} meters across. That's getting on for 10^{25} times *smaller* than the hydrogen atom, the smallest atom. Yet this lack of continuity is still a limitation, a coarseness that seems to attach to reality that doesn't exist in the imaginary mathematical world where anything can happen and continuous values are the norm. It's just possible that the universe is infinite, or that some quantum properties could correspond to real numbers (see page 191), but there is a fair chance that real numbers can only ever approximate reality.

However, Greek mathematicians were not aware of this type of issue in the physical world, nor did they spend too much effort on trying to find a form mathematics that would more accurately fit reality. Instead, they cocooned themselves in considering the isolated and perfect world of theoretical mathematics. This was a world typified by a mysterious name—Euclid.

4 Elegant Perfection

If you were exposed to geometry at school, ending with a triumphant (or irritated) QED—the contraction of *quod erat demonstrandum,* approximately "that which has been proved"—that traditionally caps a geometric proof, you were following in the footsteps of Euclid. His book *Elements* was *the* standard mathematical textbook for around 2,000 years. Yet we know so little about Euclid himself that some scholars have suggested that he never existed at all, but was instead a constructed persona for multiple authors. We do know that his work came out of the unparalleled school established by Ptolemy I at Alexandria, where Euclid (if he existed) was a key figure. His masterpiece was never intended to include earth-shattering new content. Euclid was known as a great teacher, and this was the book of the course. It was a textbook of existing knowledge, but pulled together in a very effective fashion.

The mathematics of Euclid operates in a little world of its

own, a model for the way math has operated ever since. It starts by making a few basic assumptions and then constructs a hierarchy of proofs that never require any of the dirty work—observing, counting, and measuring—that are featured in a real-world problem or in science. We think of it now as a definitive geometry textbook, but it also incorporated an introduction to arithmetic and the closest the Greeks came to algebra in their visual representations of algebraic problems—the kind of thing described in their approach to the series $1 + 1/2 + 1/4 + 1/8 \ldots$ (see page 39).

The concept of proof and exactness in mathematics, as opposed to the rough-and-ready approximation of the hands-dirty surveyor, was arguably the biggest contribution that the Greeks brought to the mathematical scene. As we have seen, the Egyptians were already active in geometry and the Babylonians had numerical and algebraic expertise that went far beyond anything the Greeks ever achieved. But neither of these earlier civilizations worried too much about proofs. For them, it was enough that the numbers or the shapes did the job. Their mathematics was anything but pure.

We have already come across the concept of mathematical proofs, because these began with the Pythagoreans, but it was Euclid who defined a more exacting model for the process. The starting point was to have a set of definitions of terms and of axioms or "givens"—things that mathematicians would need to take for granted to make their mathematics work, but that it wasn't possible to prove. Instead these axioms were (hopefully, though not always) self-evident and could be used as a starting point. The proof then required the mathematician to make a series of logical steps, one building on the other, and only using the axioms as starting points, until the proof was complete.

Euclid's book (technically books, because of the limited practical length of scrolls) begins with his definitions—what a point, a line, and a circle are, for instance—then five axioms (known as "postulates" here) and five "common notions," which we would now also consider axioms, before plunging into his first theorem, which describes how to construct an equilateral triangle with a pair of compasses and a straight edge (the Greeks were mad on constructing their geometry using just these pieces of equipment). In practice, some of Euclid's definitions were weak—he defined an angle by using a term "inclination," which is probably less familiar than the term he is defining—but he had the right idea. We won't bother with the details of his proofs here, but it is worth taking a look at those axioms.

The five postulates were:

1. A straight line can be drawn between any two points.
2. A straight line can be extended continuously.
3. A circle can be drawn for any distance from a point.
4. All right angles are equal.
5. If a straight line crosses two other straight lines and the interior angles are less than two right angles, the two other straight lines will meet.

While the five common notions were:

1. Things equal to the same thing are equal to each other.
2. If equal things are added to (other) equal things, the results are equal.
3. If equal things are taken away from (other) equal things, the results are equal.

4. Things fitted (exactly) into one another are equal.
5. The whole is greater than the part.

A lot of this may seem like obvious common sense, but the whole point about being rigorous in a mathematical proof is that you don't make unnecessary assumptions. Common sense is not an option. One of the advantages that the world of mathematics has over reality is that you can make assumptions that don't hold outside in the real world. These start in the definitions where we discover, for instance, that a line is "a length without breadth." Real lines, of course, always do have breadth. The definition immediately launches us into the world of something that is physically impossible to construct in reality.

Equally, at least one of the postulates makes assumptions about the environment in which the mathematics is undertaken—an assumption that does not appear in the axioms because Euclid never thought of it. Euclid's fifth postulate (which is a rather clumsy way of saying that nonparallel lines meet, and by implication that parallel lines don't meet) did indeed hold true for the flat surfaces on which Euclid was working, but it isn't true for a curved surface, which are much more common in the real world. Think of drawing parallel lines, for instance, at right angles to the equator of the Earth, heading north. Lines of longitude are such lines. By the time they reach the North Pole, the lines meet—even though they were genuinely parallel at the equator. The system for working on flat surfaces does not apply on curved surfaces.

A fundamental requirement for Euclid's geometry is that parallel lines do not meet. If they do, geometry as Euclid knew it would not work. It wasn't until the nineteenth century that "non-Euclidean" geometry, designed to work on real-world curved surfaces, was developed. At the time it seemed to have limited application beyond measurements on the surface of the Earth, but its extension from curved surfaces to the curvature of something more than two-dimensional would prove invaluable when Einstein began work on his general theory of relativity.

Bearing in mind the way that mathematics can exist and function without any relation to the physical universe we occupy, we have to ask if Euclid's work was about reality—and conclude that it wasn't. This is not saying it was useless. It provided a valuable approximation to reality, but the geometric world that Euclid's work occupied was not real. Probably the closest description of the relationship between Euclid's theorems and the physical objects of the real world was the metaphor developed by Plato, who flourished several hundred years earlier. Plato envisaged a hypothetical, almost heavenly, universe of perfection where mathematical structures like those that Euclid would later work on existed. These perfect shapes and objects had a kind of hyperreality, while the shapes and structures that we experience in the real world were mere shadows of the pure original.

Specifically, Plato used the image of shadows cast into a cave—if he considered, say, a drawn triangle, it was as if a true perfect triangle, out in the harsh light of true mathematical existence, was casting the fuzzy shadow of an imperfectly drawn triangle in his cave. And there is an element of truth here. Not that mathematics is somehow more real or perfect than the world, but that math can usually only approximate to the actual universe, as the universe is far too complex to model perfectly. Just as whole numbers have true real-world equivalents, but fractions don't, so the shapes of geometry fall down when faced with the physical world constructed of atoms and lines that have width and aren't perfectly straight.

A great example of the imperfection of life in the cave—of the distinction between the perfect shapes of mathematical proofs and real-life objects that we can handle and experience—is the mind-bending puzzle usually called the missing square puzzle.

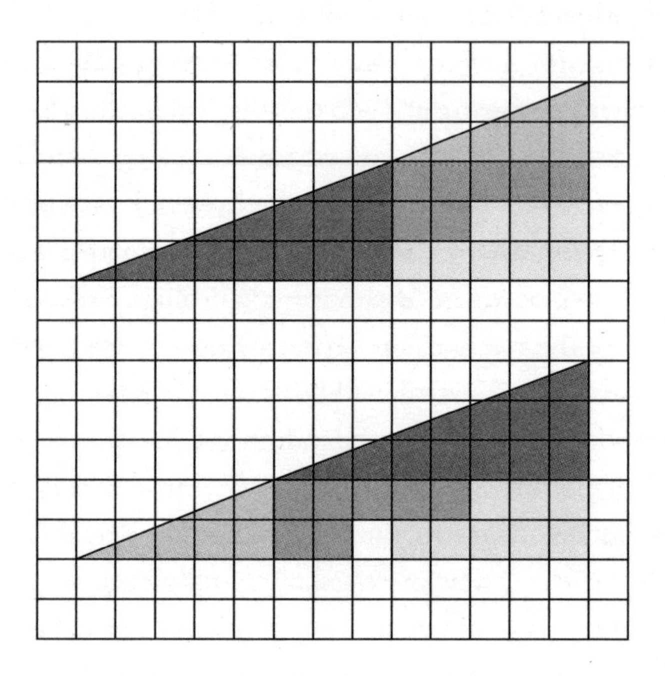

The two shapes shown above consist of exactly the same pieces, where those pieces have just been moved around on a grid. Yet the upper triangle seems to have a different area, as we are left with a blank space when the shapes are rearranged. This would not be possible with Plato's perfect shapes—but here, in the shadowy real world, we can get away with it, because the way that lines are drawn is not precise. The angle of the smaller triangle, at the left-hand side of the top image is not quite the same as the angle of the larger triangle at the top right, even though they appear to be the same. That very slight difference is enough to account for the extra bit of area that is then brought into the open by rearranging the shapes.

The further we get into this journey through the relationship between mathematics and reality, the more we will see the use of models, toy versions of reality, used by scientists. The universe is a vastly complex entity. Even a small part of it like a human being is ridiculously complex in structure. To try to understand the universe, and what makes it up in a physical sense, we usually need to simplify, to deal with less complexity than exists in the original.

Models can literally be physically constructed representations—think, for instance, of orreries, the elegant mechanical models of the solar system that were popular all the way up to the introduction of computers. However, more often than not, we use "mathematical models"—a collection of equations or mathematical systems that act to some degree like the real thing, but that are simpler to handle. Often such models will now be realized in the form of computer programs, though many scientists still prefer a model that can be reduced to (relatively) simple formulas.

We will explore these models in much depth in later chapters, but in essence, the models are reflections of parts of the universe, or even of the whole thing, that exist in Plato's idealized space. They aren't reality, but they provide a simplified, perfected equivalent to reality. They are what are sometimes called archetypes, giving us the rather beautiful medieval description of Plato's concept as being about "types and shadows." The only thing is, Plato got the relationship back to front. In his picture there were the true, real perfect versions of, say, triangles, and then there were the flawed, limited shadow version that we would encounter in the everyday world. To Plato our world was less real than the ideal one. But when scientists produce their models, the picture is inverted. It is the models that are flawed and limited. Our scientific models and theories are mathematical shadows of the natural world, much simpler than the original systems that cast those shadows.

It doesn't help that Plato's cave was, itself, a model—so arguably the cave belongs outside of the cave. You could equally say that mathematics is an agreed fiction, a shared mental world that mathematicians agree to collectively inhabit. But their world isn't allowed the looseness of literary fiction, because here all the rules have to be pinned down and agreed. As long as a piece of mathematics is consistent with those rules it is acceptable, whether or not it has any parallel with the physical universe. Yet we keep coming back to the fact that a sizable subset of mathematics not only has parallels, but has an uncanny ability to mirror what the real world can offer. It could just be because so much of the essential structure that mathematics is built on—like the nature of whole number arithmetic—started as a reflection of the real world. But whether there is anything more, we will return to once the mathematical landscape has been made clearer.

Even with its carefully proscribed world of perfect shapes, Euclidian mathematicians had their nemesis in the form of squaring the circle. With their obsession for performing everything using only a pair of compasses and a straight edge, the idea was to be able to construct a square with exactly the same area as that simplest and most perfect of shapes, a circle. Such was the obsession there was even a word in ancient Greek for the people who spent their time attempting this feat—they were τετραγονιδζειν (*tetragonidzein*). But impressive though the name may sound, they were battering their head against a brick wall.

We now know that in attempting to uncover the exact area of a circle the Greeks were coming up against an even greater irrational number than $\sqrt{2}$ in the form of pi (π). Not only is pi irrational, it is also transcendental, meaning that it isn't possible to produce a finite equation that will perfectly produce its value. (There are equations for pi, but they are all infinite series, so we can't even write down pi exactly in equation form.)

In Euclid's time, mathematicians were happy with the challenges of geometrical rigor. And it is certainly true that this wasn't math undertaken purely for intellectual stimulation. It was widely applied. As we have seen, the word "geometry" is derived from Greek words meaning measuring the Earth, and an approximate version of geometry was clearly valuable whether you were a surveyor or an architect. But all the derivations and proofs took place in the sterile, isolated mathematical universe, compared with which the real world was irritatingly messy.

The physical universe seemed an untidy corruption of the precision of Euclid's proofs. But this isolation would not last long when a mathematician entered the scene who was prepared to get his hands dirty and apply mathematics to engineering and even to the weapons of war.

5 Counting Sand

The Greek mathematician who made the mathematics of the time more practical was Archimedes—though he also succeeded in a feat of impractical number work that most at the time would have regarded impossible.

Although a fair number of texts from Ancient Greece have been recovered via later translations, many were lost forever, and among them were biographies of key figures. We know that a biography of Archimedes was written some time after his death, but it has been lost, along with the key dates in his life, so his birthdate can only be approximated to 287 BC. Archimedes proved himself just as much a wizard of geometry as his predecessors, but arguably he was also the first engineer, because the math that he employed was applied in practical ways to devise engines of war and machines for practical work.

While it was impossible to get back the purity of the imagined underlying structure linking math and the universe that the

Pythagoreans had once held, Archimedes showed that math could be the servant of the natural philosopher as a requisite for the development of science. Some people have even considered Archimedes the first scientist in the modern sense. For his own part, though, Archimedes was never more than an applied mathematician. According to the Greek historian and biographer Plutarch, Archimedes regarded his mechanical inventions as mere "diversions of geometry at play," considering them of no importance.

Most significantly for our story, Archimedes established a different kind of abstraction of mathematics. As we have seen, Greek mathematics in general and geometry in particular had been primarily visual and would never have been capable of creating the flights of mathematical fantasy that we now experience. They couldn't, apart from anything else, because the Ancient Greek number system was so limited. Greek mathematics rarely took on the symbolic forms that we are familiar with and struggled with any numerical complexity. As we have seen, the Greek numerals were clumsy—simply the letters of the alphabet with one or two old letters added to enable it to cope with the range required. And most frustratingly, the system ran out at 10,000—a myriad was the biggest number it provided.

To demonstrate that a number system could do far more, Archimedes wrote a strange short book named *The Sand-reckoner.* In it, he calculated the number of grains of sand that it would take to fill the universe. Clearly this was not one of his practical engineering feats. His intention seems to have been to demonstrate that mathematics was not limited by mundane thinking, or by the numbers as they were then known. To make it work, Archimedes had to devise a whole new number system of his own.

To give the book some authority it was addressed to the local king Gelon (or Gelo) of Syracuse. It seems that Archimedes worked for the king, and it has even been suggested that the two were related in the small melting pot that was Syracuse. Archimedes warns Gelon to avoid the mistake of thinking that even the number of grains of sand on the Earth is infinite or uncountable, and makes it clear that he refers to "not only that which exists about Syracuse and the rest of Sicily but also that which is found in every region whether inhabited or uninhabited." But then he sets the minor challenge of the Earth's actual sand aside to show that he could even produce a number to match the number of grains required to fill the universe.

In modern terms, this would be impossible to meaningfully calculate as we simply don't know how big the universe is. We have a figure for the size of the visible universe, the limit for light to travel in the lifetime of the universe, which forms a space about 90 billion light-years across, but beyond that we have no idea. In Archimedes' day, however, the picture of the universe was far simpler. By then the central fire had been abandoned for Aristotle's simpler model with the Earth at the center of everything. Everything that was seen in the sky—the Moon, the Sun, and the planets—rotated around the Earth, as they clearly appeared to do from day to day. And on the outside was a sphere containing the distant points of light that were the stars. By this reckoning, the whole thing was rather smaller than our current picture of the solar system (bearing in mind that the stars were just outside the orbit of Saturn in their model, as this was the most distant known planet).

Intriguingly, Archimedes does point out in *The Sand-reckoner* that there was an alternative viewpoint available and he quotes a book by Aristarchus of Samos, who makes the Sun and the fixed stars unmoving, with the Earth orbiting the Sun, which took over the position at the center of the universe. This is an intriguing snippet because Aristarchus's book did not survive, so we don't have the original words Aristarchus wrote on the matter and have to rely on a few indirect references to discover the first known suggestion that the Earth orbits the Sun.

Aristarchus said that his universe was far bigger than the usual model, describing the size in a way that Archimedes points out is confusing and, in truth, downright impossible. This is because Aristarchus apparently claimed that the sphere of the fixed stars was so distant that it compares with the Earth's orbit as "the center of the sphere bears to its surface." As Archimedes pointed out, since the center has no magnitude, you can hardly have a ratio based on it. However, Archimedes believed that he understood what Aristarchus really meant was a comparison between the size of the Earth—the center of the traditional universe model—and the size of the sphere of the fixed stars in the traditional model.

With this established, Archimedes set out to work out the number of grains of sand it would take to fill either the traditional universe or the larger Aristarchus version. The first requirement was to calculate how big those universes were. Since Archimedes was an Ancient Greek, this was inevitably established with a vigorous dose of geometry. The calculations were based on a number of assumptions on the size of the Earth and the relative sizes of the Earth, Moon, and Sun. Most of these were straightforward, such as "that the perimeter of the Earth is about 3,000,000 stades and not greater." (Stades were a measurement based on the distance around a stadium.) But one assumption is harder to grasp: "That the diameter of the Sun is greater than the side of the chiliagon inscribed in the greatest circle in the sphere of the universe."

It helps to know that a chiliagon (a word we don't see often enough) is a 1,000-sided polygon. Archimedes was using a very specific definition of "the sphere of the universe." He did not mean the sphere of the fixed stars, which we would tend to think of as the extremity of the universe in the Greek model, but rather "the sphere whose center is the center of the Earth and whose radius is equal to the straight line between the center of the Sun and the center of the Earth." In other words his "sphere of the universe" was actually the sphere of the apparent orbit of the Sun. So he was saying that the Sun's diameter was around 1,000th of its apparent orbit.

The popular approach to science in Ancient Greek times was one of pure argument, without much effort to confirm whether or not the results bore any resemblance to reality. However, astronomy tended to be something of an exception to this rule. Archimedes was never a pure armchair philosopher, and his father had been an active astronomer. The basis for the ratio of the Sun's size to its orbit was a spot of practical astronomical work. Archimedes got a long rod with a movable disc attached to it. Waiting until just after sunrise, when he believed the Sun's light was safe to look at (it isn't, but at least it is weakened by the extra air it has to pass through), he pointed the rod at the Sun and slid the disc up and down until it just covered the Sun's surface. From the angle the disc made to the eye, he worked out that the Sun's disc covered between 1/164th and 1/200th of a right angle, then he multiplied this up for the whole orbit and rounded up to get to his approximation of 1/1,000th of the distance around the orbit.

The distances that Archimedes used were measured in stades

(in the original Greek *stadion*, plural *stadia*), which does present us with something of a problem. As mentioned above, these were based on the distance around a stadium running track, but this unit was not standardized. It appears to have been 600 Greek feet in length, but these varied from place to place, so a stade could have been anything from around 150 to over 200 meters (490 to 650 feet) in length. Even with such a fuzzy set of definitions, though, Archimedes' calculations put the size of the universe somewhere in the outer planets, which wasn't bad given the level of estimation involved.

He then decided arbitrarily that no more than a myriad (10,000) grains of sand would fit into a poppy seed, and that a poppy seed was not less than 1/40th of a finger's breadth, the standard unit of measurement that came below a foot. All he needed now was a way to count beyond a myriad. To do this he started with a myriad myriads—100 million—and called the numbers up to this value the first order. He then multiplied a myriad myriads by itself to get the second order, multiplied the result by 100 million again to get the third order and so on. When he had done this a myriad myriad times he ended the first period and began all over again on the second period.

By adding a set of rules to perform arithmetic with his super units, Archimedes was able to calculate that the traditional universe would be filled by no more than 1,000 units of the seventh order (10^{51}, or 1 followed by 51 zeroes) grains of sand, or, for the significantly bigger universe of Aristarchus, 10 million units of the eighth order (10^{63}) grains. The number system that Archimedes had devised was not really the way forward. It built on the existing, disastrous Greek system and was clumsy at best. However, despite his use of geometry in calculating the size of the universe, there was something special and new about the way that Archimedes had worked.

Arguably, in the way he had gone through his calculations, he was making an abstraction that was pretty much unique among his Greek colleagues. He may have been dealing with grains of sand, yet his number system was one step removed from the traditional approach of being unable to separate numbers and objects. Yet despite this step forward, Archimedes' numbers still lacked an absolutely crucial component. What no one could see at the time was that the element that was truly lacking was nothing.

There was a big, zero-shaped hole.

6 The Emergence of Nothing

One of the most important reasons that the Greeks and the Romans struggled with numbers was because of a terrible absence from their arithmetic. They had no way to represent nothing—they had an absence of zero. In a sense, even those earliest accountants, scratching their tallies on bone or clay, had the concept of nothing. A clear, empty tablet had nothing on it—no stylus marks at all. I might not be able to show you no goats, but I can have an empty field that is free of goats—or an empty box that contains no oranges. Unconsciously, when we take on the concept of zero we stop thinking of a number purely as a correspondence between a marker and an object and start thinking of a container instead.

However, there is a big conceptual step from nothing as an absence of objects and a nothing that could take an active part in mathematics. Yet it was only by using zero that mathematicians could get full control of their numbers. What's more, this new wonder number could also act as a convenient placeholder, giving the way that numbers were written an enhanced structure, enabling far more complex calculations than had ever been possible before.

Zero looks like such an innocent, unimportant little thing—but it would transform mathematics. For a long time it was regarded as a special case. Many mathematicians thought that it was not really a number at all. You can see why. It misbehaves when undergoing the basic operations of arithmetic. Mathematicians (and scientists for that matter) don't like special cases, but zero is a special case par excellence. It is the only integer (whole number) that can be added to or subtracted from another number leaving it exactly as it was before. Multiply anything—*anything*—by zero and you get zero. It is a disintegrator weapon, a destroyer of other numbers.

As for dividing by zero, the process unleashed the terrifying and arithmetically useless power of infinity. Imagine dividing 10 by 1. We get 1. Now divide it by 1/2. When we learn fractions we are taught that dividing by 1/2 is the same as multiplying by 2, so the answer is 20. (If it helps, think of dividing up 10 cakes into portions of 1/2 a cake each. There will be 20 portions.) When we divide 10 by 1/4 we get 40. As the number that is being divided into the 10 gets smaller and smaller, the result gets bigger and bigger. And as the bottom number in the fraction tends to 0, the result tends to infinity.

However, the worst trick that zero has under its belt is still to come. What do you get when you divide zero by zero? Apple's voice-controlled assistant, Siri, has a very effective answer to this:

> Indeterminate: Imagine that you have zero cookies and you split them evenly among zero friends. How many cookies does

each person get? See, it doesn't make sense. Cookie Monster is sad that you have no cookies and you are sad that you have no friends.

Any fraction with zero on the top is zero, while any fraction with zero on the bottom ends up with an infinite result. Somehow zero over zero manages to be both at the same time . . . or neither. When zero first came into use in India there were major arguments among mathematicians over the outcome of this division. Over a period of times both apparent outcomes prevailed. In the seventh century, Brahmagupta was sure that zero divided by zero produced zero, while in the twelfth century Bhāskara pronounced the outcome to be infinite.

More recently, mathematicians have plumped, like Siri, for saying that the outcome is indeterminate—it doesn't have a value. Dividing zero by zero does not have a useful mathematical outcome. Mathematicians can do this because, as is becoming increasingly obvious as we move through its history, most of mathematics is not based on reality. When that is the case, we allow mathematicians to make arbitrary decisions on what the rules are going to be. Zero divided by zero is one such case.

Because of its oddities, just as mathematicians would later decide that 1 was not a prime number, they initially often considered zero not to be an integer. This caused a problem when the highly useful concept of the number line was developed. Anyone educated in the last thirty years will probably have come across the number line at school, but in case you missed out, the easiest way to imagine this is a ruler, stretching off to infinity in both directions. Each of the main markers on the ruler is an integer—so moving up the line (to the right) increases the value and moving down the line (to the left) decreases it.

The number line takes in both positive and negative numbers—so the question is, what happens when we go down below 1, heading for the negative values? One of the first practical number lines, the BC/AD dating system, fluffed this question. Devised by the monk Dionysius Exiguus in 525 and popularized by the historian the Venerable Bede in the 730s, the approach became common in Christian countries from the ninth century. In this dating system, the year before AD 1 is 1 BC. There is no year 0. And that remains the case to this day.

The line effectively jumps between –1 (1 BC) and +1 (AD 1), without anything in between. The letters AD stand for Anno Domini, "in the year of the Lord," making AD 1 the first year of Christ's life (though whether this was the intention is disputed). The gap in the number line often misleads would-be historians when trying to calculate the age of someone who crossed the BC/AD divide as, ironically, it seems likely that Jesus himself did, since most scholars now suggest that he was born around 4 BC.

This approach of jumping from –1 to 1 was just about acceptable for dating, as historians are not mathematicians, but it wasn't acceptable for the number line, because the whole point of the line as a tool is that the amount you move up and down the line corresponds to the operations of addition and subtraction. For instance, we can use the number line as a calculator to work out 5 + 2 by starting at 5 and moving up (the + direction) two places, where we find the answer is 7. But if you have a number line with no zero and try to do the operation 1 – 1, starting at 1 and moving down (the – direction) one place would take you to –1, not the correct 0. Like it or not, zero had to be an integer for the basic operations of arithmetic to work properly.

If all that zero did was to patch up the number line and represent an empty container—filling in the value that comes in the integers between –1 and 1—it would have been valuable, but not as transformative as it has actually been. This is because of zero's second role, as a placeholder in our numbers. As we've already seen (chapter 2), Roman numerals, and their Greek predecessors made it very difficult to perform calculations as there was no system to the layout of the digits that aided calculation, and because of this numbers quickly become huge unmanageable strings of characters. But in practice, the classical civilizations were ignoring a powerful technique that had already been around for more than 1,000 years.

This was back when the Babylonians used a number system of base 60 derived from the ideas of their Sumerian predecessors. Here, an upright stylus mark stood for an individual digit, just like a tally mark, and a sideways mark, called a hook, represented 10. Just as we have a day divided up into 24 hours, but then 60 minutes, the Babylonians counted up in 10s and then in 60s. Their predecessors had a separate mark for 10s from 1s, but 60s were just represented by a big version of 1, which was fine when using different styli, but could easily be confused. So by later Babylonian times an alternative approach was brought into play making use of the power of position.

When you look at a big Roman number like MDCCCLXXVII, there is a terrible waste of information. Although the Romans made a very tiny use of the position of letters, making LX 60 but XL 40, on the whole, the position was just a matter of laying things out in order. But the position of a character has the potential to carry a significant amount of information. "GOD" and "DOG" contain exactly the same characters, but the position of those characters makes a total change to the meaning. Why not do the same thing with the position of digits and make use of this extra information in a far more general way? The Babylonians did just this. If a number came in the rightmost position it represented 1s, the next position was for 60s, the next for 60×60s and so on.

So, if I use a Y for the Babylonian "1" and a D for the Babylonian "10," then the number YY DY has $2 \times 60 + 10 + 1$, so it represents 131. Not only can this be much more compact for large numbers, because we are making use of the position as an additional piece of information, it also transforms arithmetic-making operations like addition, subtraction, and multiplication vastly more easy. The reason for this is that we have similar numbers in columns, one above the other. So, for instance, to add YY DY to Y Y we can lay the sum out in columns

YY DY
 Y   Y

And quickly see that the result is YYY DYY—we've done scary base-60 arithmetic without even thinking about it. Of course you need carrying rules to get values from one column to another, just as we have now, but given that, what we have here is a vastly more effective system than that used by the Greeks and the Romans. The Babylonians even had the ability to cope with the problem of an irrational like the square root of 2 that so upset the Pythagoreans (see page 42), as remarkably they also incorporated decimals (or rather sexagesimals as they were working to base 60) into their numbering system—there is even an old Babylonian tablet that gives the square root of 2 a value that approximates to 1.414222. It's hard to understand why the Greeks took such a huge backward step when they could have learned from Babylonian methods, but any reason is now lost in the mists of time.

There was, however, one problem with the Babylonian system. In the sum above, the second number that was added to the first was Y Y—61. But how would 3,601 be represented? It has Y in the 60×60 column, nothing in the 60 column, and Y in the 1 column. So it is Y Y—the space between the Ys is technically bigger, but it is hardly obvious which of the two numbers is which, especially when producing a tablet by hand. Looking at many surviving Babylonian tablets, the layout could be extremely haphazard. Of course it's simple enough when counting those goats that I loaned to my neighbor in chapter 2. I'm hardly likely to have lent him 3,601 goats, so the context tells me what was intended. But the distinction is not at all obvious when these numbers are being used to deal with commodities that come in smaller units, or with money.

By the time the Greek civilization was taking off and the Babylonians were in the decline, a solution had been reached. If a column was empty, it was filled with a diagonal marker that identified it as a null column. So 61 was still Y Y, but 3,601 would now be represented by Y \\ Y. Brilliant. Calculations in columns were far safer and context wasn't really needed to work out what was happening. And yet \\ was not quite a modern zero. For some reason, the Babylonians never made the step of accepting that their dividing marker could go at the end of a number, so anything with zero in the units column was still in danger of being confused with other numbers. For whatever reason, \\ wasn't a fully functioning placeholder. And it never appeared on its own or in calculations, so it didn't function as the integer zero either. It was just a ghost of a zero as a part-time placeholder—

and it is only in its double role that it would be able to take on its true worth.

This same approach with an occasionally used placeholder was also used in some late Greek numbering when it cropped up in the work of the astronomers who flourished toward the end of the Ancient Greek civilization. Although conventional Greek numbers used the unwieldy system of applying a single letter to each value from 1 to 10, then to each of the multiples of 10 to 100, and each multiple of 100 to 1,000 there were also systems that came closer to the Babylonian approach. For example, in later days, angles were measured as we do in degrees, minutes, and seconds. To represent a value of 5 degrees, 0 minutes, and 20 seconds, the Greeks would put an empty placeholder in the minutes and even in the seconds, a void that was represented by a circle with a complex bar shape over it.

They had made a small step closer to zero in some respects then, but this placeholder symbol was still not treated separately as one of the numbers. Where ordinary numbers were represented by the appropriate Greek letter with a straight bar over it, the empty placeholder was marked differently to indicate this was something odd, outside the normal. It might seem to us that it was such an obvious step forward to turn it into an actual number, but despite the inevitable use of numbers in accounting and those astronomical texts, we have to remember that the vast majority of Greek mathematics was visual and concerned with shapes; it remained geometrical. And with that mind-set, an empty shape was hard to envisage, and even harder to make any use of in practical geometry.

This meant that the mathematicians and philosophers who might sensibly have come up with the mathematical concept of zero never had a starting to point to work with. As for those accountants and merchants who needed to work with numbers, the Greeks generally used a form of wireless abacus (actually the source of the name "abacus")—a board with a set of columns in which pebbles could be placed. This was little more than an advanced version of my goat accounting using fingers. There was no need for a placeholder, because the fixed columns of the counting board automatically put the 10s and units and so on into appropriate columns. When a section of the board was empty, that was fine—but it wasn't in any sense a number, just a gap in this advanced form of tally.

It wasn't until the early thirteenth century that zero in its modern form reached Western mathematicians at the hands of several mathematicians, notably Leonardo of Pisa, better known by his filial nickname Fibonacci (it's a contraction of "filius Bonaccii"—son of Bonaccio). His father was a diplomat who seems to have taken young Fibonacci, born in 1170, with him on his trips representing Pisa in North Africa, and it was on these journeys that Fibonacci may have become acquainted with the novel numbering system that the Arab mathematicians had improved from an Indian original. In his book *Liber Abaci,* which translates as "Book of the Abacus," a rather odd title as it had nothing to with the abacus, Fibonacci both introduced an early version of the now standard "Arabic" numerals to the West, and brought us zephirum, a word that probably translated from the Arabic *sifr,* a special kind of number represented by a simple upright egg-shaped figure, 0—the zero.

Those powerfully flexible numbers originally from India could have reached the West far sooner. In AD 662 a Syrian bishop named Severus Sebokht pointed out that the "Hindus" had made "subtle discoveries in astronomy" and specifically highlighted "their valuable methods of calculation, and their computing that surpasses description. I wish only to say that this computation is done by means of nine signs." Sebokht's reference to the effectiveness of the nine numerals other than zero was a response to a parochial view of Greek scholars who seemed to believe that there was no wisdom outside of Greece. Unfortunately it wasn't enough to gain the system any attention. Again, the French mathematician Gerbert, later Pope Sylvester II, appears to have briefly taught Arab/Indian numerals without the zero toward the end of the tenth century, but once more it seems not to have caught on.

Although the exact origins of the modern true mathematical zero is fuzzy, to say the least, the assumption is that Arab mathematicians got the concept from working with the highly advanced mathematicians of India. In his book *Finding Zero,* mathematician and science writer Amir Aczel describes undertaking a lengthy mission to try to find a missing example of zero that predated Arabic use. His hope was to show for certain that this was neither a European nor an Arab invention. The French-Hungarian scholar George Cœdès, working in the 1920s, had translated a Cambodian inscription from a ruined temple that dated the stone to the 605th year of the çaka era, which we know to have begun in AD 78, putting the date of the inscription as AD 683, before other well-dated recorded use of zero.

The significance of this inscription is that the number 605 used a zero (represented by a dot) as a placeholder. It was a self-dating early use of zero (though of itself it doesn't show the true zero in action, just the equivalent of the much older Babylonian \\ placeholder). This discovery was written up in the *Bulletin of the School of Oriental and African Studies* in 1931. But the inscription itself was lost and had not been photographed. The evidence was dependent on the reliability of the reporting by

Cœdès. Later an almost-as-old zero, dating to AD 684, would turn up in Sumatra, but Aczel was determined to track down what he believed to be the oldest known example.

After a long search, Aczel discovered the stone fragment K-127 in Siem Reap in a shed full of stone inscriptions that had survived the destruction by the Khmer Rouge, the brutal Communist party that ruled Cambodia in the late 1970s. There was the inscription he had been searching for, complete with that enigmatic 605. To the untrained eye the dot for the zero does look rather like a later disfigurement, a simple scraped out space, but this was the evidence of that tantalizingly early use of zero. However, despite Aczel's enthusiasm, there seems to be no evidence that this was any more than the kind of place marker used by the Babylonians and the true evolution of the full numeric zero seems to have come via a different route.

There is good evidence that early Indian mathematics and astronomy was influenced by Greek sources, and it seems likely that the placeholder circle used by Ptolemy and others found its way to India from there—but then it was in India in the minds of the superbly creative mathematicians of the period that the breakthrough was made, with zero allowed to perform its full numerical role. Indian mathematicians had been using the concept of "nothing," an absence of number since at least the sixth century, but we don't know exactly when the zero digit came into play. Confusingly, the same term was used at the time both for the placeholder of an empty column and for an unknown quantity—the kind of thing algebra has taught us to represent as "x, the unknown."

Arguably this equivalence of the placeholder and the unknown arises because each represents a kind of empty space waiting to be filled. We can even see this coming through in one of our words for zero, null or nil, from the French, "null," which originates in the Latin phrase *nulla figura,* meaning "no number." Although there are many nearly-zeroes all the way through from ancient Greek to Arab mathematics, the oldest certain Indian example of a placeholder zero is in a stone tablet from Gwalior dated AD 876 that uses a small circle for zero in both 270 and 50. But it's very likely that it was in use before then.

Zero as the numerical value you get when you take a number from itself was certainly firmly in place for the great Indian mathematician Brahmagupta to make use of it in the seventh century. It seems likely then that the fully functional zero we now use had origins both in Babylonian ideas, via the Greeks, and in the Far East before becoming more widely used in India. From there it reached the growing world of Arab science and mathematics to begin a new stage in the increasing sophistication of numbers. It was via this route that zero would finally arrive (along with the Indian number system) in Europe, which is why we still refer to our numbers as "Arabic numerals" rather than the more accurate term Indian.

If this were a formal history of mathematics it would be important to consider the way math was used in China and South and Central America, and particularly to explore more the work of the early Indian mathematicians, but as we are concerned with mathematics as it interfaces with science, the use of zero and Indian numerals represents by far the biggest contribution to the development of modern science. Probably the next most important contribution from India was interesting in that it had an extra level of abstraction from reality over most of the mathematics of the time.

This was the concept of the sine in trigonometry, the mathematics of the angles and lines of a triangle—the name means triangle measurement. The Greeks had used a messy system to deal with this called a table of chords, but the Indians introduced the modern concept of the sine of an angle, which, if it escapes you from school days, is the ratio of the length of the side opposite an angle to the size of the side opposite the right angle in a right-angled triangle. You may well have forgotten this, which, in a sense, reflects the existence of that extra level of abstraction. The side of a triangle or an angle is something tangible, but a sine is a particular ratio, which is meaningless unless you know how it is derived. Sines are very useful, but are arguably not as real as the components from which they were produced.

Zero, though, was a vastly bigger and more powerful concept than the sine, and when introduced to the West by Fibonacci it was to a mixed reception. Mathematicians seem to have taken to the power that zero could provide rather more enthusiastically than the public at large first did. This certainly appears to be the case if we take seriously the poet John Donne's complaint in a 1620s sermon "The less anything is, the less we know it: how invisible, how unintelligible a thing, then, is this nothing!"

The negative reaction to the new numbering system was not entirely an emotional one. Accountants noticed the ease with which a 0 could be converted into a 6 or a 9 in the new number system. This was considered enough of an enticement to fraud that in 1299 the Florence city council published an edict that numerals should not be used in accounts, where numbers had to always be represented by words to avoid tampering. Even in the time of Galileo, a Belgian cleric had to warn his suppliers that they were only to use words for numbers in their contracts.

But before the flourishing of mathematics in Europe inspired by the new notation and the power of zero, a new group of mathematical scholars in the Middle East took the Indian notation and ran with it. As we have seen, the system reached the West from the Middle East, and so became known as Arabic numerals, but the key book that transformed the way mathematics worked was titled *On the Calculation with Hindu Numerals* by its Persian author al-Khwarizmi around AD 825. This was translated into Latin as *Algoritmi de numero Indorum,* and it was the Latinized version of al-Khwarizmi's name that gave us the term "algorithm" for the kind of step-by-step rules that a computer typically performs. Another book by the same author brought us the word "algebra." Now, thanks to the flexibility of the new number system, math was able to take a step away into a new kind of reality that paralleled many possible physical worlds.

Algebra might be a subject of fear for many school students, but it has an unrivalled power in its puzzle-like problem-solving ability and its open-ended approach. It's useful to think back to the Ancient Greek way of doing things to see why the Indian numerals and concepts like algebra were so powerful. As soon as the Greeks moved away from their elegant geometry they struggled because their approach to fractions was limited and difficult to use, and because they had no mechanism to deal with the kind of problem that we would deal with trivially using algebra. When they moved away from diagrams and visual thinking they simply didn't have the tools to deal with the mathematics.

Take a trivially simple equation like $A+B=C+D$. The Greeks had no way of dealing with a process like this symbolically. The best they could do is write out the whole thing in words, a process not helped by the convention of the time of not leaving spaces between words, so an approximation to a Greek style equivalent would be

THEAANDTHEBTAKENTOGETHERAREEQU
ALTOTHECANDTHEDTAKENTOGETHER

In reality it would be worse than that, as the Greeks also had no tradition of replacing an unknown quantity by a letter or other simple symbol (it would have been highly confusing if they had used letters, because the letters were already being used for numbers), so the "word equation" would not have A, B, C, and D, but rather words that reflected the actual things being combined. This was another example of the backward step that resulted in not taking up Babylonian techniques—the different attitude the Babylonians had to numbers enabled them to take on algebra-style problems, even dealing with a form of quadratic equations in a way that would be forgotten for millennia.

Inevitably, Greek mathematics did move on, and by the time of the late Alexandrians from around AD 250, approaches to algebra had reached an intermediate state, typified by the work of Diophantus. He seems to have reached back to the more number-oriented approach of the Babylonians, but adding in the Greek essential of accuracy and proof, rather than approximation. Perhaps the biggest contribution Diophantus made was in developing a symbolic approach to algebra problems that made them far more compact and easier to address.

In his book *Arithmetica,* Diophantus made use of symbols and structures to indicate unknown values, powers of 10, and arithmetic operations. What he produced was not a true equation as we would now use one, but a symbolic representation of such an equation that was consistent and compact. So, for instance, his equivalent of the equation $2x^4 + 3x^3 - 4x^2 + 5x - 6$, if we use our lettering with S for square, C for cube, x for unknown, M for minus, and u for unit, would have been SS2 C3 x5 M S4 u6.

Diophantus provided the starting point for the flourishing of algebra.

In a sense, al-Khwarizmi's book on algebra, *Al-jabr wa'l muqābalah* (the meaning is uncertain), was a reversal of some of the work Diophantus did, in that it worked purely in words without even his clumsy equation representations, but al-Khwarizmi presented something much closer to a modern basic algebra primer in the way that he worked through the solving of equations, particularly quadratic equations—even if it was just in word form. Some sources have suggested that the development of algebra was driven by the need to calculate the outcome of the complicated rules for inheritance in the Arabic world of the period—which might also explain why al-Khwarizmi generally avoided negative solutions to his equations.

With the translation of Arabic works like al-Khwarizmi's, things begun to turn around. As we saw, the new numbering system was resisted by the general public for centuries after Fibonacci's *Liber Abaci* was published, but this was more the last struggles of a dying system than any sensible operation. Those who had a vested interest because of their expertise in using the abacus or a counting table (like an abacus, but without the wires) may have resisted the new numbers as opening their field of expertise up to the common herd—ironically, given the name of Fibonacci's book. And we all grumble about "new math" that isn't done the way we were taught at school. But the change could not be resisted. The superiority of the new system was so obvious to anyone who understood the workings of mathematics that there could be little doubt that it would take over.

In the medieval period, many scholars played down the importance of mathematics. But one individual would begin a crusade to ensure its significance was recognized.

He Who Is Ignorant

7

Mention the name "Bacon" to most scientists and they will almost certainly think of the sixteenth-century politician and philosopher who developed an early version of the scientific method, Francis Bacon. But Francis was pretty much indifferent to mathematics. For him science was all about collecting, codifying, and classifying. His was the kind of activity that the great physicist Ernest Rutherford would later mock by saying "All science is either physics or stamp collecting." However, Roger Bacon, apparently no relation, gave the impression that he had a much more modern take on things. "He who is ignorant of mathematics," wrote Roger, "cannot know the other sciences and the things of this world." Bacon wanted numbers and mathematics to have a status in understanding nature that was simply not recognized at the time.

Bacon appears to have been a fascinating character, though tracking down anything definitive about him as an individual

800 years after his birth has proven difficult. Apart from some unreliable paperwork dating from a century after he died, the main source we have is Bacon's own writing and he rarely gives anything more than tantalizing glimpses of biographical detail. Even Bacon's date of birth, usually given as 1214, is only accessible indirectly via a mechanism that involves the unusual need to incorporate a value judgment into a calculation.

In 1267, in his masterpiece the *Opus Majus,* Bacon wrote: "I have labored diligently in sciences and languages, and forty years have passed since I first learned the alphabet . . . for all but two of those forty years I have been in study." The assumption is usually made that the forty years he spent *in studio* (as he put it in the original Latin) began when he went up to Oxford University, and as matriculation, the formal entry into a university, was then usually at the age of thirteen, this takes us back to 1214. But that would make Bacon's rather cryptic remark about having "first learned the alphabet" metaphorical, referring to the point where he took on serious study. If it were literally the point when he first learned the alphabet, then it makes his birthdate later—perhaps around 1220. But 1214 remains the most widely accepted figure.

According to the fifteenth-century historian John Rous, Bacon was born in the sleepy country town of Ilchester, though there is no other source (certainly Bacon never mentions where he came from) to back this up. What we do know, though, is that Bacon went up to Oxford University, where his first task was likely to get his hair cut. Before they could attend university, students had to undertake minor orders, effectively becoming junior monks. One essential for this was to take the tonsure, the haircut that leaves a large bald patch on the top of the head. One of the earliest businesses to open in Oxford after the university was established was a barbershop. At the time none of the colleges we would now find in Oxford had been founded and there were no university buildings in the modern sense. Accommodation was in lodging houses, while lectures took place in large rooms scattered around the city, wherever space could be rented by the teachers.

For that matter, the university itself bore little resemblance to the kind of institution of that name familiar now around the world. Oxford started when a schoolmaster called Theobald of Étampes decided to set up a school for higher education in the town in 1095. This wasn't an obvious location for Britain's first university—there were plenty of cathedral cities with a tradition of the kind of ecclesiastical structures required, and the abbeys were already providing centers of learning that could easily have been extended. But Oxford had proved strategically significant in the civil war that raged across England between King Stephen and the Empress Matilda in the twelfth century and it was the transport links left over from its central position that probably made it the focus for the UK's first university.

Just as the establishment was getting settled in, modeling itself on the University of Paris, which began teaching in the mid-twelfth century, a murder in 1209 threatened to tear Oxford apart. The mistress of one of the (supposedly celibate) students was murdered and angry mobs of townspeople, straight out of central casting, were soon baying for student blood. The town officials reacted with summary justice by hanging two other students who simply happened to be in the wrong place at the wrong time.

The university authorities weren't sure what was worse—that the townspeople should hang innocent students, or that the locals concerned themselves with matters that the university considered to be within its own jurisdiction. Seventy of the university's masters, a fair percentage of the teaching staff, left the town for a smaller group of schools that had recently been established in an East Anglian backwater called Cambridge. Thankfully for Oxford, a visit from the papal legate, Cardinal Nicholas, in England to sort out the recalcitrant King John, restored order to the town. In 1214, the basic structures of the university were established, though it would not formally receive its charter until 1231.

When Roger Bacon arrived, perhaps in 1227, the university was still a very new establishment. Although it was technically part of the church hierarchy, this didn't stop it from being a lively, even a dangerous place. Attacks leading to death in the gutter were commonplace, either through criminal assault or becoming caught up in the large-scale brawls that regularly broke out between townsfolk and members of the university. In their dress, the scholars would not have been much different from the youths of the town, but the characteristic tonsure made sure that they stood out in a crowd. There was also fierce rivalry between different student factions from the north and south of the country.

A good example of the atmosphere around the time that Bacon was studying at Oxford is found in a report from 1238. The new papal legate, Otto, was visiting the town and staying at Osney Abbey, located nearby to the southwest. A party of students and masters went out to the abbey to greet the legate, but the pleasant formality went horribly wrong when the master cook of the establishment threw a pot of boiling water over an unfortunate Irish student who was begging at the abbey door. One of the other students then drew his bow and shot the cook dead (even on a social occasion, appropriate weaponry was considered necessary and acceptable for protection).

The welcoming party turned into a seething mob, and the legate had to be smuggled to safety farther down the Thames at Wallingford. There was a drastic clampdown on university freedoms, only lifted several months later once the regent masters, the ruling lecturers at the university, had performed an act of penance, walking barefoot through the streets of London to the legate's residence. Bacon, who was painfully plain speaking and incapable of the kind of political activities necessary to make it up the academic greasy pole, was unlikely to have been a regent master and so would not have taken part in this degrading display of subservience. As for the students, they soon resumed their wild behavior.

It would have taken Bacon six years to receive his BA (bear in mind that the first part of medieval university training was essentially a high school education) and then another two to become a master, Magister Artis, an MA licensed to teach in the university. He could then have spent eight years more to become a master of theology, with eight further years of studious work before he could take the only available doctorate in theology, but there is no evidence that he did so. Instead Bacon moved to Paris, where there was much demand for Oxford masters, as Aristotle's work had been banned at the French university, but was now creeping back into the curriculum without any local experts left to teach it. Paris was generally regarded as the greatest of the three universities widely recognized in the West at this time: Oxford, Cambridge, and Paris.

Bacon only stayed a few years, but in Paris he met Peter of Maricourt, a mysterious figure known as Peter Peregrinus (roughly "wandering Peter"). Peter, who wrote one of the earliest surviving treatises on magnetism, steered Bacon toward an interest in the sciences, inspiring a clear enthusiasm to undertake experiments. Returning to Oxford in the late 1240s, Bacon spent serious sums of money on books and kit. He wrote in 1267: "During the twenty years in which I have labored specially in the study of wisdom, after abandoning the usual methods, I have spent more than £2,000 on secret books and languages and instruments and mathematical tables etc." At the time, a substantial house would have cost 2 to 3 pounds to build. While there may have been a degree of exaggeration in this claim, Bacon certainly seems to have worked his way through the family fortune funding his scientific endeavors—and mathematics was already an important part of his toolkit for understanding the world.

Running out of money may well have been the reason why, soon after his return to England, Bacon joined the newly formed religious order, the Franciscans. These gray-robed friars were building a large establishment in Oxford that would stretch from the city's south gate to the castle in the west. Friars had more freedom than the longer-established monastics like the Order of St. Benedict. What's more, at Oxford the house had tight links with the university. Bacon would have had access to books and was able to spend a considerable amount of his time on natural philosophy. All seemed to go well until around 1250, when Bacon was dispatched to the mother convent in Paris and says that he spent ten years on "menial tasks." It is entirely possible that the outspoken Bacon upset the authorities one too many times with his opinions.

While in Paris, Bacon developed ideas on calendar reform, realizing that the Julian calendar then in use and dating back to

Roman times made the year around eleven minutes too long. He calculated that this meant that the calendar went a day out of synchronization with the seasons every 125 to 130 years (the actual figure is 128 years). This doesn't sound much of a shift, but since the introduction of the calendar it had slipped more than a week from its original dating, something that Bacon found intolerable, as it meant that religious festivals would be held on the wrong dates. His plea for reform was ignored and it was not until 1582 that the Gregorian calendar, almost identical to the approach that Bacon detailed, was adopted in Catholic countries, not reaching Britain until 1752. Different parts of the United States typically converted with the countries that had originally colonized them, leaving dating confused for decades.

While Bacon was still in Paris, a new head of the Franciscans was elected to return the order to its roots of poverty and lack of possessions. He began to shift the gray friars away from their academic links, eventually banning them from writing books. But Bacon had become obsessed with communicating science and searched for ways that he could get around the ban. He wrote to a number of influential people, including Cardinal Guy de Foulques, the papal legate to England. Bacon asked the cardinal for special permission to be excluded from the ban so that he could write a book on science, but the message got garbled and de Foulques replied by asking to see this (nonexistent) book immediately.

What Bacon had hoped for was sponsorship—with his family finances exhausted and no support from the Franciscans, he needed an external source of money for research and materials. Instead what he received from de Foulques were demands. While Bacon was nervously wondering how to respond to the cardinal, he received unexpected news. In 1264 de Foulques was summoned to Perugia to discover that the College of Cardinals had elected him pope, and the next year he was crowned Clement IV. Suddenly Bacon's friend in high places had reached the top of the heap.

In 1266, a letter arrived from Clement, ordering Bacon to get started on his exploration of science, and giving him authority to ignore the prohibitions of his order. Bacon still needed resources, though, and tried to raise money from his friends, only procuring a small amount. By January 1267, he decided instead to send the pope a brief proposal to procure further funding and rushed into writing with what would become his masterpiece, the *Opus Majus*. Anyone who has written a nonfiction book knows just what a painful process producing an initial proposal can be. It requires the author to squeeze the contents of an entire book into brief summaries.

Bacon could not manage the necessary restraint. His "brief" proposal ended up as a massive book over 500,000 words long (around six times the length of *Are Numbers Real?*), covering optics, astronomy, mechanics, alchemy, agriculture, and medicine, with a final section that concentrated on experimental science. While the *Opus* was being copied, because everything at the time was handwritten, Bacon decided a covering letter was necessary as the proposal had got out of hand, but this became so long it was soon a book in its own right. Astonishingly, this happened a third time, resulting in a third volume, so his final package for the pope was a three-book epic of around a million words, all written in twelve months. (In practice, only two books were sent, as the third was still being copied when the messenger left.)

By now, Bacon had begun pulling together resources in Oxford, ready for the eagerly anticipated go-ahead to write his masterpiece. But the response he received from the continent left him in despair. While his proposal was still on the road, Clement had died. With little remaining support in the hierarchy, Bacon lashed out against those in the church who thought it was enough to know theology and who were ignorant of science. According to a chronicle written in 1370, Bacon was condemned for "suspected novelties." He may have been imprisoned at Ancona in central Italy and seems to have been released by 1290, recorded by Rous as dying in 1292.

Bacon was no great mathematician himself, but he left behind in his *Opus Majus* an incomparable picture of the understanding of the world at the time. The most unusual focus he had was on the importance of mathematics for natural philosophers (and indeed for theologians). Along with his intellectual predecessor Robert Grosseteste who also emphasized the essential nature of mathematics, Bacon was certainly setting the direction for the future, just as he did with calendar reform, putting him in an interesting position that could be called the grandfather effect.

We are used to individuals being labeled the "father of X" where X is some new approach—think "the father of computing" or "the father of space flight," but there are also grandfather figures who have, to my mind, a subtly different role. The grandfather figures set a broad direction often with a dead-end approach, but did not actually come up with a viable means to head in that direction. Some time later, the father figures take the first significant step that is in direct line of inheritance to the finished product.

I'd suggest that the Victorian photographer Eadweard Muybridge was the grandfather of moving pictures. (His real name was Edward Muggeridge, but he was always flamboyant and regularly reinvented himself.) In his work, first at the behest of railroad magnate Leland Stanford in California, and then at the University of Pennsylvania, Muybridge produced many thousands of series of images of motion, captured by using a bank of cameras triggered one after the other in sequence as horses, people, and animals passed by. As an exercise in its own right, producing these series of still photographs was useful and informative, but what Muybridge also did was to display the images using a crude projector that produced true, if very short, moving pictures. He even had constructed the first purpose-built movie theater at the World's Columbian Exposition on the outskirts of Chicago in 1893.

Muybridge's improbably named Zoopraxographical Hall was a full-scale dedicated building in which he gave educational lectures on his work—but in practice what the audiences came in to see for their 25 cents was boxers fighting and scantily clad women parading up and down on the big screen. (Muybridge preferred to capture images of humans with little or no clothing, even producing sequences depicting naked bricklayers in action, so that it was possible to study their musculature.) Although the quality of Muybridge's photographic work has always been recognized, his moving pictures were long ignored, in part because of a concerted campaign by a historian of film to have his work forgotten. It's certainly true that Muybridge's technique was a dead end, where the mainstream technology needed the use of roll film, but there is equally no doubt he demonstrated the concept and earned that "grandfather" status.

Like Muybridge, Bacon did not make any great strides in the advancement of the field, in his case mathematics, but he did emphasize its importance at a time when most academics had a clear disdain for the subject. Bacon may well have had an inferiority complex because, at the time, the only subject in which it was possible to achieve a doctorate was theology. But, he pointed out, the theologians of the day were dismissive of the importance of mathematics, not understanding how valuable it could be. It seems in part that this was due to an uncomfortable association that math had with magic.

Bacon himself was strident in his attack on magic, showing that charlatans used trickery to prey on uneducated people and get money out of them. Yet many at the time, even academics, confused magic and mathematics. This was due to an unfortunate coincidence that the word pronounced "matesis," meaning scientific knowledge and the entirely different word "mathesis" meaning divination were both written as "mathesis." It was also the case that as late as Tudor times, "calculating" was used to mean undertaking magic and it was common for the uninitiated to confuse mathematical and magical texts.

There is no doubt that Roger Bacon did not have great expertise in mathematics. In his *Opus Majus,* for instance, he makes the blunder of thinking that a good example of Aristotle's lack of complete knowledge was that the Ancient Greek philosopher confessed to be ignorant about squaring the circle, something that Bacon says is "a problem that is clearly understood in these days." As we have seen (see page 52), squaring the circle or producing a square of the same area as a circle using only the geometer's tools of compasses and straight edge, was a problem that taxed mathematicians since ancient times, but that ultimately proved impossible.

Despite his limited skill, though, Bacon was without doubt remarkable for his time in emphasizing the importance of mathematics. The British mathematician and contemporary of Isaac Newton, John Wallis wrote to Newton's mathematical nemesis (see page 110), Gottfried Leibniz toward the end of the seventeenth century:

> Those who in the present century (following Galileo) joined mathematics to natural philosophy have advanced physics to an enormous extent. This was also being attempted by Roger Bacon (a great man in a dark century) four hundred years ago (and more).

Bacon dedicated a whole section of his *Opus Majus* to mathematics, though he used the term to cover everything from calendar reform to astrology. This last may seem unnerving to those brought up to consider astrology totally unscientific, but Bacon was a person of his time. While he dismissed astrology when used as a form of fortune-telling, putting in the same realm as other fraudulent attempts to extract money from easily deceived members of the public, he did believe that the arrangement of the heavens at the time of a person's birth could influence their personality, and it was in this form—a kind of early nature-versus-nurture debate—that astrology was seen as potentially a respectable science in his day.

Bacon and his contemporaries had no way of knowing how little influence the planets could have on humans, but this idea of astrology as potentially giving a guide to personality (a bit like the Myers-Briggs and other personality-type indicators beloved of some corporate businesses today) made sense to those employing it, given the state of scientific knowledge of the day. Looking at the more general need for mathematics in science, Bacon pointed out its benefits:

> Of [the] sciences the gate and the key is mathematics . . . He who is ignorant of mathematics cannot know the other sciences and the things of this world. . . . Moreover, what is worse, men who are ignorant of mathematics do not perceive their igno-

> rance, and therefore seek no remedy. While, on the other hand, knowledge of this science prepares the mind and elevates it to a sure knowledge of all things, so that if it perceives these roots of wisdom which surround this science and applies these roots to an inquiry into other sciences and things, then it will be able to know all things in sequence without doubt or error, and in ease and power.

For Bacon, mathematics was more than a tool, it provided a way of thinking, a structured application of logic that those ignorant of math did understand, but that made the human mind better able to address and to understand nature. He went on to demonstrate the importance of mathematics, for instance, in understanding optics (the scientific subject that he studied in most detail, and in which he had both original theories and experiments), which requires a good understanding of angles, geometry, and symmetry.

It is not that Bacon was the only one who had spotted that mathematics should be taught. (The education of the period had some mathematical content. The "quadrivium," a significant section of the university curriculum, consisted of arithmetic, geometry, astronomy, and music—all either directly mathematical or with mathematical elements at their core.) It was rather that Bacon was different in seeing the importance of mathematics to all of science, and in taking it out of its confined world and suggesting that it provided a mechanism for understanding.

Largely speaking, Bacon's view would be ignored for a couple of centuries. However, there were a few lights in the metaphorical mathematical darkness. One, also from Oxford, came in the form of the Oxford Calculators, also known as the Merton School or the Merton Mathematicians. There were a handful of men associated with Merton College in Oxford, one of the first Oxford colleges, who took the importance of mathematics (and advancing it) seriously.

The group, notably Thomas Bradwardine and William Heytesbury worked on logic, geometry, and methods of calculation, but are probably best known for devising the law of falling bodies, or the "mean speed theorem" that says that if you have an object undergoing constant acceleration (such as a falling body) and one going at a constant speed, they will cover the same distance if the velocity of the constant speed body is half that of the final speed of the body that is accelerating. Interesting in its own right, perhaps the most important significance of the work of Bradwardine and his colleagues was their move away from Aristotle's ideas on motion to a position that was partway to that of Galileo and Newton. Most importantly, unlike Aristotle's drive from pure philosophy, for the Merton calculators, mathematics had far more significance in producing results.

Another example of the ability to extend mathematical thinking came from the fourteenth-century French scholar, Nicole Oresme. It was he, for instance, who worked out what we would now consider the mathematics of powers—so, for instance, that $y^2 \times y^3 = y^5$. When we multiply two items that are raised to a power like this, we add the powers together. Similarly he envisaged the possibility of what we would now think of as fractional powers, such as $x^{½}$—though the concept of a power as we now use it did not explicitly exist, so this was an indirect approach, but he was still starting to make more of an abstraction from reality than the work at Merton, which dealt with squares and cubes only in reference to the observed behavior of moving objects.

Less abstract, but equally important, Oresme also seems to

have devised the idea of using a graphical representation of a mathematical structure. Today we take it for granted that it can be useful to draw a chart representing, say the function $y=x^2$, which will enable us to get a better visual understanding of that function, and also allows us to investigate concepts like differentiation and integration that we will meet with the introduction of calculus (see page 111). Oresme at the very least was the first to popularize this approach, then called the latitude of forms, and even speculated about the possibility of extending the approach into three dimensions.

By insisting on the importance of math and of experiment, Bacon and his medieval successors bridged the natural philosophy of his day and the true science of the future. For Bacon, math was a tool to help us to understand the universe, and as such could never exist in true isolation from it. It's probably just as well that he never came across the concept of imaginary and complex numbers.

8 All In the Imagination

An imaginary number, the name mathematicians give to the square root of a negative number, is a concept that at first seems totally alien to the way that numbers had been related to the real world up to this point—so weird, in fact, that it appears to prove the final detachment of mathematics from any vestige of a connection to the reality. Yet over time, imaginary numbers have proved surprisingly versatile in helping us understand the physical, and are employed every day in calculations by down-to-earth engineers.

It all started with the negative numbers. Once mathematicians had got their heads around these variants on the counting numbers as a practical concept, they put them through their mathematical paces. It turned out that multiplying two negative numbers together produces a positive one. It might not seem entirely obvious why this is the case, but the easiest way to think of it is to use the number line (see page 65). The minus sign effectively represents a change of direction on the line, so by taking a pair of changes away from positive, you are left heading in the positive direction once more. Another way of thinking of it is that it was one of the mathematicians' arbitrary decisions, but one they had to make, as exploring the alternatives shows that it's the only option that results in consistent results elsewhere.

The problem then was that if multiplying a negative number by itself produced a positive result, and we already know that multiplying a *positive* number by itself also produces a positive result, what was the square root of a negative number? What number, when multiplied by itself, came out negative? It couldn't be either a positive or negative number, which seems to leave few options.

Before we get too far down that line of thinking, it's worth considering the relationship between negative numbers and reality a little further. As we saw in chapter 2, a negative number is useful in bookkeeping as a way of representing a debt, and it can be regarded as a mechanism to indicate how many goats have been taken away when we end up with fewer than we started with—but these examples are really still about positive numbers, used in a very specific way. I can't show you a debt, for instance, I can only show you the money that I will use to repay it. So can I produce something of the real world that is explicitly negative? As it happens I can, provided it is acceptable to go beyond pure counting. But this was not realized as a possibility until the nineteenth century.

Think of the labels on the poles of an electrical battery. One is positive, one negative. This description is generally attributed to Benjamin Franklin, though when first applied it really simply meant something different and didn't have the true mathematical distinction of a positive and negative number. However we now know that the charge on, say, an electron and a proton are equal and opposite. They don't have a direction, so it's not like the equal and opposite force of Newton's third law—they are purely numerical ("scalar" as mathematicians and physicists would say) in nature. Yet the two operate just like positive and negative numbers in the way that they add up and cancel out.

Although the definition of which charge is positive and which is negative is arbitrary, the charges are effectively physical objects that operate as positive and negative numbers respectively. It might seem at first sight that modern discoveries also produce examples of real-world rational numbers. We now believe that particles like protons and neutrons are made up of smaller particles called quarks, which have charges of 2/3 or −1/3. However, although these are genuine exact values, not approximations, they only appear to be rational numbers because we didn't know about them when we assigned charge values of 1 to a proton and −1 to an electron. Really, quarks have charges of 2 or −1, while protons and electrons have charges of 3 and −3 respectively.

Returning to the square roots of negative numbers, once it was realized that they could not be produced by any existing type of number, mathematicians made the arbitrary decision to make them up. This wasn't because there was a demand for them, but simply the mathematician's urge to boldly explore new numerical worlds. Given the appropriate name "imaginary numbers" by a sarcastic René Descartes, these strange creatures became the new playmates of the mathematician. It was like opening up a new dimension in the world of math, one that seemed to have no equivalent in the physical world. Imaginary numbers were initially pure mathematical toys—but surprisingly flexible ones because they extended all kinds of mathematical operation.

The basic idea of an imaginary number emerged in a work by the Milanese physician and mathematician Girolamo Cardano, dating back to the first half of the sixteenth century. Cardano's *Artis Magnae Sive de Regulis Algebraicis Liber Unus* (book one of the great art of algebraic rules) is probably best known for a degree of treachery involved in writing it. Cardano had learned from fellow mathematician and engineer Niccolo Tartaglia a method for solving cubic equations (an equation with a cubed unknown, and potentially lower powers as well, such as $x^3+4x^2+2x+5=0$) on the assurance that Cardano would never tell anyone about the method. But Cardano published it in his book. This wasn't plagiarism—he fully acknowledged Tartaglia's role—but he had broken his word.

However, in passing in the book, Cardano also commented on the solution to an apparently harmless little equation like $x^2+1=0$. As this is the same as $x^2=-1$ (produced by subtracting 1 from both sides of the equal sign), its solution requires there to be a number that, multiplied by itself, produces a negative value. Cardano remarked that such a number was "as subtle as it is useless," an assessment that would prove far from the truth. The useful part would not come until the nineteenth century, though, when the German mathematician Carl Friedrich Gauss realized that imaginary numbers made it easy to extend the number line to turn it into a two-dimensional number plane.

As we've already seen, it is common to think of the integers stretching along a horizontal line with 0 in the middle, a line that heads off toward negative infinity on the left and positive infinity on the right. Gauss put a second number line at right angles to the first, with positive imaginary numbers heading upward and negative imaginary numbers going downward. Now, any point on the plane could be defined using a single "complex number"—a value combining a real number and an imaginary one. With the square root of -1 written as i, a complex number might be $5 + 2i$—this would define the point that is five units along the real (horizontal) axis and two units up the imaginary (vertical) axis.

It might seem that this is nothing more than relabeling the x and y coordinates we are familiar with on charts, but the power of the complex number is that it was possible to perform algebra on it in exactly the same way we would a normal number and all the rules and mechanisms still applied, but produced a result that worked in this two-dimensional space. Such complex numbers proved ideal for describing phenomena like waves that are naturally two-dimensional in form—and so imaginary factors crept into everything from basic electrical calculations to sophisticated quantum mechanical equations. As long as any final result shed the imaginary part, so you didn't end up with, say, an imaginary electrical current, these versatile numbers proved and remain a very powerful mathematical tool.

Do imaginary numbers exist? They certainly don't have a direct correspondence to anything in the physical world. I can't, for instance, show you 3i apples. I can't even use the kind of indirect method that's possible to show you -3 apples when I take three away for an existing set of apples. However, anything can exist in the mathematical universe as long as it is properly defined and consistent with the rules. And in this case, because the way that the imaginary (and particularly complex) numbers behave to usefully model two-dimensional changes, they have proved excellent tools for coming to a real-world solution via an excursion into the abstract mathematical world.

Imaginary numbers provided a tool that doesn't exist in the real world. We can export a problem from the real world to the imaginary universe, operate on it in a way that wouldn't work without the availability of imaginary numbers, then translate it back to reality. Where a simple counting number can be considered to have a direct, one-to-one correspondence to an object or a set of objects, imaginary and complex numbers operate in a parallel universe, but are still able to give us insights into physical reality.

Before imaginary numbers really took off practically, though, old-fashioned, straightforward positive numbers and geometry would be used to conquer the universe.

9 The Amazing Mechanical Mathematical Universe

One of the reasons it took a long time for math to play its full part in science was that the generally held view of the universe was riddled with metaphysical mysteries. When natural philosophers believed that everything outside the Moon's orbit was perfect and made of a different element to the rest of creation (the so-called quintessence), and perhaps even rotated because it was powered by angels, it was hard to see how mathematics could be of any help. Yet Galileo opened cracks in the ancient Greek model of the universe, and made the first significant steps toward the use of math to predict the behavior, for instance, of projectiles and pendulums.

Riding on the shoulders of this giant, Isaac Newton constructed a mechanical picture of the universe, where a good enough mathematician, with perfect data and immense calculating power, could know everything. Although Newton was too much a Christian (if an unorthodox one) to say this, his successor and superfan, the eighteenth-century French scholar Pierre-Simon, Marquis de Laplace, had no such hesitation. Laplace was unusual for his day in being an atheist. (According to legend, when Napoleon asked Laplace of God's place in his philosophy he said, "I had no need of that hypothesis.") And he believed that the universe was literally a hugely complex mechanism like a vast clock, where with the right information and intelligence it would be possible to perfectly predict the future. He wrote:

> Given for one instant an intelligence which could comprehend all the forces by which nature is animated and the respective situation of the beings who compose it—an intelligence sufficiently vast to submit these data to analysis—it would embrace in the same formula the movements of the greatest bodies of the universe and those of the lightest atom; for it, nothing would be uncertain and the future, as the past, would be present to its eyes.

Newton would never have gone this far, yet emphasizing his position on the cusp of the change in the fundamental place of mathematics in science, Newton made the calculations that would enable him to establish his laws of motion and gravity in a new, mysterious mathematics, the method of fluxions, that dealt with the infinitesimally small. His mathematics might not have predicted the future perfectly, but it gave a remarkable ability to predict forces and their impact, particularly the mysterious force of gravitation. Yet to avoid scaring his audience, or more likely to make his methods less obvious, Newton painstakingly translated as much of his work as he could into old-fashioned geometry for his masterpiece, the *Principia*. Newton's clockwork universe was one of certainty and predictability.

Newton's translation was possible thanks to the work of his French predecessor, the philosopher René Descartes. We tend to remember Descartes for two things—proclaiming "I think, therefore I am," and having the so-called "Cartesian coordinates" named after him, where we specify locations on a chart with x and y values. But this was just a tiny part of his work, which took in everything from theories on light to attempts to take a scientific view on the soul.

Those Cartesian coordinates were far more than a simple way of representing a point as a pair of numbers (an approach that Bacon was well aware of in the thirteenth century). Descartes was responsible for analytical geometry, a mechanism for translating from geometrical forms to the equivalent algebraic equations and vice versa. For example, an equation like $y = x^2 + 2x + 3$ could be represented by plotting out the values as x/y coordinates on a chart. Similarly, many natural processes that were known to change in a way that could be plotted out on a chart could be matched to an algebraic equation, making it much easier to predict outcomes.

This approach of Descartes, fulfilling Oresme's concept of plotting a function as a chart (see page 93) would prove immensely powerful in the development of Newton's work and enabled him to hide away most of his algebraic workings in the form of less transparent geometry. Descartes himself seems to have been unaware of the power of the idea. He says in *La Géométrie*, where he introduced the concept, that what he was providing was an easy mechanism for constructing geometric forms—in essence, he had in mind more the way that Newton would turn algebra into geometry than what we now see as the real power of turning spatial problems into algebra. But whatever his intent, Descartes provided a mechanism for translating the geometry that had a more visual link to the world around us to the more abstract-feeling algebra that was the ancestor of most modern scientific mathematics.

Newton's mathematical wizardry, the method of fluxions, had a long ancestry. It had been realized as far back as Ancient Greek times that by slicing up a shape into smaller and smaller segments it was possible to make an approximate calculation of the area of that shape. For instance, if you want to find the area of a circle, you could imagine dividing that circle up with a series of straight cuts from the center to the circumference, producing segments like two-dimensional equivalents of orange segments. If you make these segments thinner and thinner, they become closer and closer to triangles, for which the area is easy to calculate. Pile up those segments in alternating directions and you have something that approximates to a rectangle that is r wide and πr high. It doesn't take a mathematical genius to work out what the area of the circle is.

Although approaches along this line had been suggested since the time of the Ancient Greeks, it was the fifteenth-century German philosopher Nicholas of Cusa who made use of the method to come up with the familiar πr^2. He didn't suggest that this was a proper mathematical approach that produced an accurate result as that would mean dealing with an infinite set of infinitely small segments, but Nicholas accepted that the method effectively predicted what the right answer would be as the segments got thinner and thinner, and the result of piling up the segments became closer and closer to a true rectangle.

Others took up this approach, notably the astronomer Johannes Kepler, but it was Newton's contemporary John Wallis, a mathematician who would have been far more famous had he not been overshadowed by Newton, who suggested a way to get around the problem of dealing with the infinitely small. He suggested that the small segments being used to, for instance, find a total area, could be considered "dilutable"—that they be made smaller to the extent that was required, without ever totally disappearing away. The term is distinctly suggestive of the approach taken by Newton in his great leap forward, developing his method of fluxions where he was always referring to flowing quantities (the very name, fluxions, suggests this). And Newton's fluxions would not just enable him to crack the problems of understanding gravity—they would start a mathematicians' war that lasted 100 years.

Isaac Newton was a man with an extraordinary breadth of interest. This is clear from a look at the catalog of his library. By the time of his death he had around 2,100 books—a substantial collection at the time, over half the size of the entire library at his college, Trinity College, Cambridge. However, alongside his 235 books on the physical sciences and mathematics, there were 138 on alchemy and a massive 477 on theology. His interests didn't stop there either, with 207 works of literature, 46 travel books, 31 on economics, and even 6 titles on medals. (He would become Master of the Royal Mint, the organization responsible for Britain's coins and medals.) Such was his diversity of focus that for much of his working life he spent significantly more time on alchemy and theology than he did on science.

Even so, Newton's contributions to science, from a better understanding of light and color to his theory of universal gravitation, were enormous. And underpinning his tour de force on gravitation and motion, the book *Philosophiæ Naturalis Principia Mathematica,* usually shortened to the *Principia,* was the method of fluxions with its powerful mechanism. At its heart was an unnatural-feeling mathematical tool that was brought into play in predicting the behavior of every component part of the universe.

It is part of the Newton myth that was already coalescing toward the end of his life, when he had become the first celebrity scientist, that Newton devised the method of fluxions in his early twenties. Back then he had an enforced break at home on his family farm in Lincolnshire, when Cambridge was evacuated because of an outbreak of the plague. And he certainly gave some thought to the matter then. But from his notebooks it seems clear that Newton gradually assembled his thoughts on fluxions over a couple of decades, while he would not publish full details until a considerable time after he had developed it.

Although this new way of doing mathematics would also deal with the method of calculating areas by dividing a shape into increasingly thin slices that we have already seen, Newton's method of fluxions was primarily assembled to calculate the outcome of factors that changed with time like acceleration, essential to deal with the force of gravity, and to compare, for instance, how the Moon travels around the Earth with the way that an apple falls. We can see how the method of fluxions works by performing a quick calculation on acceleration using Newton's approach. This will involve a few equations. They are surprisingly painless, but feel free to skip them if they aren't for you.

Acceleration is the rate at which velocity (speed and direction) changes over time. To keep things simple in this example, we'll assume that the direction of movement stays the same, so we're only dealing with the rate at which the speed increases. This is easy to work out in a steady "linear" relationship where speed is, for instance, 10 miles per hour after 1 second, 20 miles per hour after 2 seconds, 30 miles per hour after 3 seconds and so on. To work out the acceleration we can say that it is the change in speed per second—in this case the speed changes by 10 miles per hour every second. A convenient way of looking at this acceleration is that it's like the gradient of a hill. Take a look at a chart of how speed changes with time:

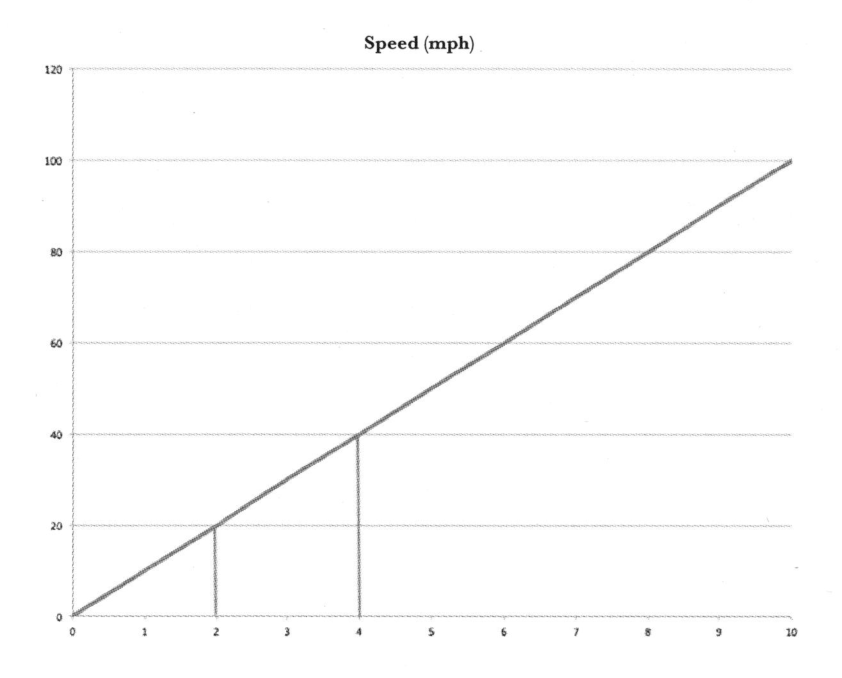

The acceleration is just the slope or gradient of that line—it is the change in speed, divided by the change in time. But in the real world, a lot of relationships don't turn out to be as simple as straight lines. Newton knew from reasonably early on, for instance, that the force of gravity was an inverse square law—one that varied with the square of the distance away from the source. The result of plotting how speed varies with this kind of relationship is not a straight line, but rather a curve.

Let's take a look at the acceleration where there is a simple square relationship between speed and time. The acceleration is such that speed = time^2. In that case, the relationship between speed and time would plot out like this:

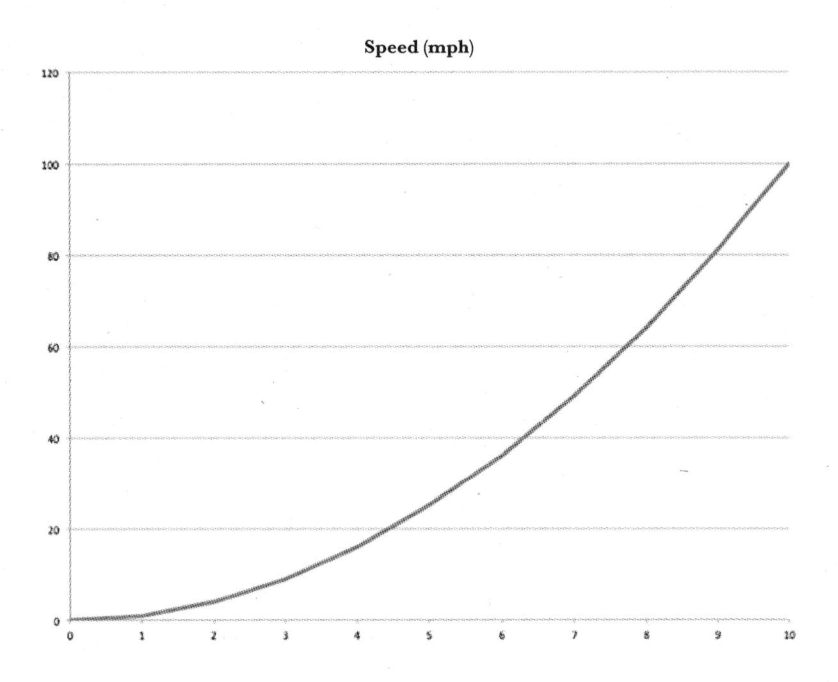

Because the result is not a convenient straight line, we can no longer simply divide the change in speed by the change in time. However if we considered a very short moment in time, then the curve will *almost* be straight, and we can approximate to the outcome during that moment by using the same old approach of dividing the change in speed by the change in time. This is what Newton did. Let's call speed s and time t. Newton called the teensy extra bit of time that goes by in the short moment a "fluxion," represented by a sort of squashed zero, *o*. So the change in time when we roll things forward a tiny bit is $(t + o) - t$, which makes the change in speed, bearing mind that $s = t^2$:

$$(t + o)^2 - t^2$$

So to get our acceleration, we divide the change in speed by the change in time:

$$\frac{(t+o)^2-t^2}{(t+o)-t}$$

If we multiply out the squares and get rid of the brackets, this becomes:

$$\frac{t^2+to+ot+o^2-t^2}{t+o-t}$$

Which simplifies to:

$$\frac{2to+o^2}{o}$$

Now we can cancel *o* top and bottom and get:

$$2t+o$$

Finally we let the little fluxion *o* flow away to nothing and we get the answer 2t, which is correct. We have worked out that the acceleration at time t is 2t. However, to get to this correct answer, something worryingly dubious has happened. When *o* becomes zero, then the previous step where *o* was canceled out top and bottom involves dividing zero by zero—which as we have seen is not considered an acceptable mathematical operation.

Newton was well aware of this problem and tried to deal with it by what amounted to concealment. He said that he only ever dealt with ratios, and that his fluxion was, as John Wallis had said, dilutable—a flowing substance that disappeared away, but wasn't actually zero. It wasn't a great argument, but the fact was that the method of fluxions worked and Newton wasn't going to abandon it just because it wasn't entirely mathematically sound. He managed to conceal as much as possible in the *Principia* by translating his algebraic calculations into geometrical arguments whenever he could, but he would soon have a more pressing problem on his hands: competition.

This came from the German mathematician, Gottfried Wilhelm Leibniz. Independently from Newton, as far as we can tell, Leibniz had been working on his own equivalent of the method of fluxions. Newton was aware that there was some work underway in Germany because both he and Leibniz corresponded with the Royal Society in London. Newton exchanged a couple of suspicious letters with Leibniz, notably one where he made use of one of the conventions of the day and wrote:

> I cannot proceed with the explanation of the fluxions now, I have preferred to conceal it thus: 6accdæ13eff7i3l9n4o4qrr4s8t 12vx

What Newton had done was to write down a sentence he felt summarized the method, then noted the frequency of the letters in the sentence and used this as a coded note of his achievement. The idea was that this would establish Newton's priority, though the sentence encoded above—*Data æquatione quotcunque fluentes quantitates involvente, fluxiones invenire: et vice versa* (given an equation that consists of any number of flowing quantities, to find the fluxions: and vice versa)—hardly clarifies what his method involved. (He did provide further cryptic clarification later on.) At this point Newton could easily have published his method and established his priority, but throughout his life he was reluctant to share his ideas, often having them wheedled out of him by colleagues. He kept his workings to himself.

Any doubts Newton had about his competitor came to a head in 1684, when Leibniz published the details of his equivalent to the method of fluxions, which he called calculus, after the Latin name for the little stones used in classical calculating tables. Leibniz used a significantly different notation than Newton, which proved considerably easier to use, but the principles were the same. Where Newton would indicate the rate of change of, say, x by putting a dot over it, a style known as pricked notation, Leibniz picked up on the convention of using the Greek letter delta to indicate a small change and transformed this to the letter d to indicate the infinitesimal change that Newton called a fluxion. This made Leibniz's rate of change be represented as dx/dt—both clearer (it was easy to miss Newton's dot) and more flexible in approach.

Leibniz also had a distinct notation for the reverse of this "differential calculus," used for the approach of adding up disappearingly thin slices to calculate an area that Nicholas of Cusa had championed. Leibniz called this integration and used an elongated S symbol, ∫, to represent "summa," the Latin for a sum. Newton had no real cause to complain. It had been his choice not to publish his method. What he did, though, was gradually to build a weight of feeling among British mathematicians that Leibniz was guilty of plagiarism.

Eventually, in 1708, the Scottish mathematician John Keill made this accusation explicit (quite possibly at Newton's request) in the *Philosophical Transactions of the Royal Society.* As tensions rose, the Royal Society of which both Newton and Leibniz were fellows, announced an inquiry, setting up a commission of eleven men to establish who had priority. The report, written by the president of the Royal Society himself, found in Newton's favor—though this is probably no surprise given that the president at the time was Isaac Newton. The outcome was a frosty relationship between British and continental mathematicians that lasted for decades to come.

Whichever approach was adopted among the mathematical fraternity, and whoever got there first, both men would come under fire from the philosopher bishop George Berkeley who pointed it out in a magnificently titled paper, *The Analyst: A Discourse Addressed to an Infidel Mathematician,* that something was astray in the whole concept. The infidel Berkeley was addressing seems to have been the astronomer Edmond Halley, who had been instrumental in getting Newton's *Principia* published. Halley was an atheist and challenged Berkeley's beliefs. In response, Berkeley tore into the method of fluxions.

He pointed out that fluxions were being used for undertaking calculations even though they had effectively flowed away to nothing—he referred to them poetically as "the ghosts of departed quantities." Using such an approach seemed to depend on faith to accept an unimaginable concept. Berkeley considered this hypocritical in those who criticized religion because it made exactly the same demand of its followers. Whatever his motives, the bishop was correct in highlighting the problem that we have already discovered. It was true that the method of fluxions (and, for that matter, calculus) worked by performing arithmetical operations with what were effectively zeroes, procedures that weren't mathematically viable.

According to Berkeley, Newton and Leibniz only arrived at the correct result by chance, as a result of two errors canceling each other out. As he put it, "by virtue of a twofold mistake you arrive, though not at science, yet at the truth." Newton's escape route depended on the fluid nature of his small changes—he would argue that they were in the act of flowing away, rather than being absolutely gone. In the example above, when we go from 2t + *o* to 2t, Newton would say that the result *tended* to 2t as *o* tended to zero, while never actually making the infinitesimally small value become zero. This fix would remain mathematical hand-waving without a firm basis until the nineteenth century when two mathematicians would finally patch up calculus for good.

In the 1820s, Augustin-Louis Cauchy effectively redefined infinity and the infinitesimally small, for the purposes of calculus, as being variable—just a label for something that tended toward a value. Then in the 1850s, Karl Weierstrass introduced the concept of limits, the standard approach used today, which makes it possible to establish a final value as the limit of making something infinitesimally small if the result approaches that limit faster than a required minimum. Weierstrass provided a formal mechanism for proving that calculus really did work as long as limits were being approached quickly enough. In a sense, Weierstrass did away with infinity in the workings of calculus. His new version of potential infinity never required that final result to be approached, as long as the (finite) rate of getting closer was big enough.

We will return to infinity in a big way in chapter 12, but for a while here it seemed that something that was difficult, perhaps even impossible, to conceive in the real world had influenced mathematics, proving extremely valuable in making Newton's vision of a mechanical universe real. Although the establishment of limits meant that infinity never truly came into play, calculus feels like mathematics that somehow manages to sneak up on reality by making use of a magical world of the infinitesimal and infinite. And yet, paradoxically, it only achieves a match with reality when this is done.

At the heart of this paradox is that calculus requires us to imagine what is happening in an instant of time—yet in a truly instantaneous moment it seems impossible that anything can happen. This was reflected in another of the paradoxes of the Greek philosopher Zeno (see page 37), known as the arrow. Although it's not the exact wording of the original paradox, probably the best way to envisage the arrow in action is to imagine that somehow we have an arrow floating stationary in front of us in space, and another arrow flashes past it, fired from a bow.

Let's imagine freezing time and taking a look at the two arrows at the moment the second arrow is exactly alongside the first. In that frozen moment in time, the two arrows appear identical. One is moving; the other isn't, but in that moment, both hang in space. Our inability to distinguish between the two, Zeno suggested, shows how artificial our ideas of motion and change are.

We now would say that there are clear differences in some of the physical properties of the two arrows. The moving arrow has inertia. Without time moving on we can't detect this, but it is still there. For that matter, special relativity (see page 211) makes it clear that anything moving has extra mass, so the arrows would be distinguishable by making an exact comparison of their masses, had that been possible to the Greeks.

Although the image of the arrows is puzzling, it also establishes that, despite its odd feel, calculus, and the way it is applied to understanding nature, is grounded in the real world. Provided we can cope with concepts like an instant in time or an infinitesimal change in position, then calculus is the natural mathematical approach to take. It was derived, after all, not from obscure mathematical considerations but from understanding how things change in nature while taking smaller and smaller views on them.

All this was very much about the behavior of Newton's clockwork universe, which would inspire Laplace's vision where the future was fully determined by the past, provided we had enough information. However, by the time Newton did his work, the seeds were already being sown to enable mathematics to take a far less certain view of the future—a view based on chance and circumstance.

10 The Mystery of "Maybe"

Statistics seems a harmless enough concept, related straight back to those prehistoric finger counts based on goats transactions. If I kept a set of tallies from different times that my neighbor borrowed goats, I could compare them and see how my neighbor's goat-borrowing habit ebbed and flowed with time. I wouldn't be able to do anything more than make direct comparisons of higher and lower levels, but I would be able to indulge in basic statistics.

Originally derived as a word from the same source as "state," the discipline of statistics started as little more than an accumulation of data on a country, the kind of thing you would find in the CIA's *World Factbook* today. This was surely a harmless enough activity—and yet the famous quote, "There are three kinds of lies: lies, damned lies, and statistics," reflects a viewpoint that seems to have persisted as long as statisticians have existed. They might claim to be innocent mathematicians, but they are clearly up to no good.

Exactly who was being quoted on the matter of lies and statistics is not clear. The quotation is often said to have been the work of the British prime minister Benjamin Disraeli, who was known for his witty and sarcastic turns of phrase. He denied responsibility, claiming that he was himself quoting Mark Twain—yet no one has found evidence of these words in Twain's written work. Perhaps it was a passing remark to the statesman when the author was visiting Britain.

It's certainly true that from its earliest days, statistics has, as a discipline, had something of a macabre feel to it. The first statistician worked on the subject of death. His name was John Graunt, and he was no professional mathematician, but rather a button maker who had an interest in how the world worked. Graunt got hold of the "bills of mortality," tables giving details of deaths in London from 1604 to 1661 and added in what he could find about births, then pulled these together in a book where he attempted to give a picture of the underbelly of life in London through the study of numbers.

In part this was a matter of presenting the existing data more usefully, collecting together numbers that had been spread across many different documents so, for the first time, for instance, it was possible to see how deaths from the plague went up and down from year to year. However, Graunt wasn't satisfied with simply reformatting existing data. He also combined numbers in a way that produced information that hadn't existed previously. So, for instance, he made early estimates of the population of London (this was before there was a regular census) based on secondary data, and also tried to produce a picture of how different life expectancies would be spread around a group of people.

It was this life-expectancy work, plus some later analysis by the astronomer Edmond Halley, that led directly to the founding of an industry that could be said to prey on human uncertainty in the combination of statistics and what might happen in the future—the insurance business. Starting in the coffee houses of London, which were then bustling meeting places for business, this large-scale regime of betting on future outcomes based on the best current statistics spread across the world to become an inevitability of life.

Despite Disraeli's disdain, statistics as a discipline on its own seems fairly staid. Yet when statistics and probability—the mathematics of chance—came together, sparks flew. If there had been any conception of the relation of mathematics and the world to date, it had been one of math slavishly modeling what was happening now, or explaining what had happened in the past. But this new brand of math, championed by lowlifes, dared to describe the future. This was not the clockwork future of Newton and Laplace's universe, but a future of uncertainty and risk. The result was a major extension of the reach of mathematics in describing aspects of the universe that were yet to be—and eventually probability and statistics would become essential to describing everything from the behavior of gases to the mysterious quantum, as we will discover in chapter 13.

To succeed in the blossoming insurance industry, it wasn't enough to have good numbers. They had to be turned, as Graunt had shown was possible, into a crystal ball that would help predict the future. And this meant turning to the habits of a far lower class of beings than mere button makers. This was the world of the gambler. In the end, the insurance business is a form of casino where the house, in the form of the industry, hopes to stay ahead by offering odds to the customers that give the "players" the chance to get back more than they put in, but in most cases will result in profits for the insurance company.

Gambling has been in existence for a long time. Polished knucklebones called astragli, an early form of four-sided dice, have been discovered in archaeological sites dating back thousands of years. And as long as coins have existed, coin-tossing games seem to have proliferated, using the head and tail sides of the coin as a simple generator of random values—at least, random as long as the coin is honest. Human beings have been happy to bet on everything from races to the weather for as long as records go back. However, gambling was largely a matter of intuition and guesswork for both player and for the honest game organizer until the Italian mathematician (and enthusiastic gambler) Girolamo Cardano came on the scene.

We've already met Cardano in the world of imaginary numbers and mathematical intrigue (see page 98), but his attempts to put the vagaries of chance onto a mathematical footing would be just as important, both to the future of mathematics, and to its separation from the everyday and tangible. Cardano was born around 1500 and wrote his book on what would become known as probability before he was thirty, though it wasn't finished until he was in his sixties and wouldn't be published until the 1660s, by which time it should have been old news. The fact that it wasn't shows how advanced Cardano's thinking was. The book was *Liber de Ludo Aleae* (*Book of Games of Chance*).

It would have been impossible for anyone who played many coin-tossing games not to notice that, with a fair coin, each side had the same chance of turning up. There was no way of knowing what the next toss would reveal, but neither head nor tail was more likely to be thrown. What Cardano did was to take this simple, practical observation and turn it into a numerical structure—to wed together the numerical concept of fractions with a view to what might happen in the future, providing an insight into the workings of a simple, easy-to-understand system like a coin toss.

Of course, that proviso about using a fair coin was an important one. One of the difficulties with making probability respectable was the frequency with which gamblers, particularly professional gamblers, cheated. Whether they were armed with a double-headed coin to win the toss whenever they chose, or were masters of the simple but highly deceptive "flick from the top" technique* used in the ubiquitous three-card trick (also known as "find the lady"), the professionals seemed almost magically able to mislead the poor marks who took part in their games. The borderline between professional gambler, illusionist, and thief was fuzzy at best.

When I give talks about probability and statistics, I usually open with a coin-tossing example. I produce a coin, which I tell the audience that I have spent some time before the event tossing until I managed to toss nine heads in a row. (This is perfectly feasible, though it does usually take some time.) I then ask the audience if my next toss is more likely to be a tail, after I have thrown so many heads, if it has a 50/50 outcome, or if it is more likely to be a head, because this coin clearly prefers heads. A few will always say "more likely to be a tail," a viewpoint that

* In the three-card trick, the operator has a stack of three cards in one hand and appears to lay them out by dropping a card from the bottom of the stack. With practice, though, especially if the cards are slightly bent, it's possible to make the card come from the top, which is how the trick is worked.

is called the gamblers' fallacy, because in reality the coin has no memory. It can't react to what has happened before. Yet it's very difficult after a run of a particular outcome not to assume that the opposite outcome is more likely.

Most of the audience will usually give the correct mathematical answer of 50/50. But a few do go for heads again. This could be just a matter of another fallacy, the "hot hand" fallacy that often crops up in sports. This is where fans assume that a run of good results means that an individual or team is on a "winning streak." However, by the time I have tossed three more heads in a row, the audience is beginning to be suspicious. And they are right to be: I am using a double-headed coin. (At this point the question is always, "Where did you get it?" It was eBay.) What's interesting is that there is inevitably a fascination with the coin, the same kind of fascination we feel for clever con tricks in the movies. The audience wants to see the double-headed coin, to get their hands on this wicked prop.

So back in Cardano's day there was awareness that either side of a *fair* coin had the same chance of coming up. (This isn't strictly true. Because of the way we flip them, a standard coin has a slightly higher chance of ending up showing whatever side is upward to start with—it's about a 51/49 chance.) But this feeling of equal chances had not been put in a form that enabled mathematics to be engaged. Although we have various different ways of expressing the chance of, say, a head coming up in a coin toss—an even chance, for instance, or 50/50—the most useful way to make progress mathematically is to use a number that we can manipulate with arithmetic. It was Cardano who was the first to represent probabilities as ranging from 0—meaning "it won't happen"—through to 1 for "it definitely will happen." This made the chances of heads turning up on a toss a 1/2.

This is fairly straightforward, but it was just the starting point of Cardano's efforts to put guesses of future chance onto a sound mathematical footing. (Cardano wouldn't have used the term "probability," which had been in use meaning "uncertain but likely" from the fourteenth century in France, but the first recorded use of the modern mathematical sense of "probability" only dates to 1692.) Taking the same approach as he had with the coin, it was now possible to say that the chance of picking a particular card out of a regular modern deck, for instance, would be 1/52.

Cardano also worked out two of the most essential ways of combining probabilities that would prove essential to any games player (bear in mind that he was both a mathematician and a fanatical gambler). The first method made possible the task of combining the probabilities of getting several possible outcomes. So, for instance, from Cardano's initial insight we know that there is a 1/6 chance of getting any particular number—a six, say—with a single throw of a die. But if you want to know the chance of getting either a one or a six, then the outcome is going to be 2/6 or 1/3.

Similarly, Cardano was able to show that to get the same throw on each of two dice—the chance of getting a double six or of getting "snake eyes" in craps—the mechanism of combining the probabilities is to multiply the fractions together. This makes the chance $1/6 \times 1/6$ or 1/36. There is just 1 chance in 36 that the specific double required will come up. What's more, he realized that this was subtly different from wanting to get a one and a six from two dice. In that case, the desired outcome could be produced by getting a one on the first die and a six on the other, or a six on the first and a one on the other, making the chance 2/36 or 1/18.

However, the cleverest realization Cardano had was to find the way to work out what the chance of getting a six with *either* of two dice (or any other multiple combination of number generators). The situation where, for instance, I've got a throw of two dice and I need to get at least one six—I don't care how. This kind of combination of probability is one we face all the time. The natural inclination is to do some sort of addition. There's a 1/6 chance of getting a six on each die, so a first shot might be just to add these together. But this clearly is wrong. If this were true, you would only have to throw six dice to guarantee getting a six, which anyone who has played dice will realize is not the case.

The problem was to find a way to represent the "on either die" selection. Cardano's genius was to notice that you could turn the same problem into a "on both dice" requirement, which could then be handled by the technique he had already developed of multiplying the probabilities together. If the chance of getting a six with one die was 1/6, then the chance of *not* getting a six was 5/6. So the chance of not getting a six with *both* dice was $5/6 \times 5/6$ or 25/36. Which meant that the chance of getting a six with either was $1 - 25/36$ or 11/36. Notice that this is just a little less than twice the chance of getting a six with a single die, which would be 12/36. And as the number of dice thrown goes up, the value gets closer and closer to 1 and certainty, but it never quite makes it, always leaving a tiny margin for having a whole lot of throws with no six coming up.

Cardano's work was built on by those who followed him, notably the French mathematicians Blaise Pascal and Pierre de Fermat, who between them solved a notorious problem that would make probability the tool of preference for the insurance trade. The puzzle, known as the problem of points, featured a game in which two equally matched players are playing for an amount of money. The game is structured so that the first person to reach a certain number of points gets the prize. But they have to stop playing before the game is finished. How do they divide up the prize money?

Let's imagine players get a point when they win a game, and at the time they have to stop, one player has 12 points and the other 7. To work out how to make the split, Pascal thought that the important thing to consider was how many rounds each player would have to succeed in, if the game had continued until it had an overall winner. Say the target to win was 15 points. In this case, the first player only needed success in 3 more games to win, while the second player needed 8. By looking at what was likely to happen as a result of the distance away from winning, Pascal was able to produce a mathematical statement of the fair way to split the prize. What he had developed, effectively, was a concept called expected value, the result that is expected after a large number of runs of a process with random outcomes.

To take a very simple example of expected value, imagine you were asked to play a game where you threw a die ten times and won the average of the numbers thrown in dollars. How much would it be worth paying to take part in the game? Common sense says that your winnings are likely to be in the middle of the possibilities and for once common sense (not usually the best guide when it comes to dealing with probabilities) is right. If you don't give any thought to it, you might say that value would be 3, as it's half of 6. But if you think about putting the values 1 to 6 in a row, the middle of the range sits between 3 and 4, so the expected value is 3.5.

You could work this out more formally by thinking you have a 1/6 chance of getting 1, a 1/6 chance of getting 2 and so on, up to a 1 in 6 chance of getting 6. So by adding $1 \times 1/6 + 2 \times 1/6 +$

$3\times 1/6 \ldots 6\times 1/6$ you would get 21/6 or 3.5. So if the expected amount you could win is $3.50, it should be worth paying anything less than that to take part. On any particular game you might lose. But as long as you play the game a sufficient number of times (and you have enough stake money) you should win overall.

This concept of calculating the expected value of a transaction is the basis for a whole range of modern financial systems—and it is by no means limited to gambling. Most notably, insurance companies are like the player in the game. They try to set the odds so that, though they might lose one particular "game" (or "policy" as they like to call them), the insurance company wins overall. This is also true, of course, of casinos. Importantly, the approach can be used to compare different options, to see which is the most attractive.

So, for instance, imagine that you were offered two possible investments. One has a 1/2 probability of returning $1,000 and a 1/2 probability of returning nothing. The other has a 1/4 probability of returning $1,900 and a 3/4 probability of returning nothing. Which is better? We can work out the expected value from the probability times the outcome. So in the first case, the expected value is $500 and in the second case it is $475, making the first the better investment, even though there is a potential to win more in the second. You can also add together different possible outcomes if there is more than one possibility for a particular investment.

The expected value, like other probability-based mechanisms for predicting the future, is not magic. It can't do the impossible. It won't tell you what you will win on a single roll of a die. But it does give a picture of the likely outcome as long as the process is repeated a good number of times. At least, when playing the right kind of game. As a member of the extremely talented Bernoulli family pointed out, there are some circumstances where the concept of expected value is not a useful guide.

Before we get to the Bernoulli demonstration, we can see the way that expected value isn't always appropriate in a simplistic way with a rather silly imaginary lottery. (I am keeping the examples to games of chance, where the probabilities can be calculated accurately, but the same approach can be taken with business investments, insurance deals, and so forth, where instead we have to rely on the best guess for a probability.)

In this lottery there are two tickets for sale, each costing $10. One has a 9/10 chance of winning $11.11. The other has a 1/100,000 chance of winning $1 million. In each case, the expected value is $10. To have an expected value that is the same as the entry price is an unusually good outcome for a lottery. In gambling games like a lottery or casino game, the expected value usually has to be less than the cost of the ticket, so that the organizer is likely to make a profit. But my lottery is especially generous. Because the expected value is the same in each case, we should be indifferent as to which of the tickets we are allocated. But they feel like very different outcomes—and the outcome that appeals most is likely not to depend on the expected value, but on your personal circumstances. The choice of which ticket to buy would probably depend on how important $10 is in your daily life.

Let me illustrate this effect with a more dramatic example. In talks I give based on my book *Dice World*, I run a psychological test called the ultimatum game with the audience. This game is much used by psychologists to show up how economists don't understand people. (Psychologists love to show up the failings of economists.) In the usual way that the game is played, a small prize, such as $1, is made available to the two players. The first player tells the other player how this money will be split between them, and then the second player says "Yes" or "No." If the second player says "Yes," the money is split as the first player announced it would be. If the second player says "No," then no one gets anything.

Economists and logicians would assume that the second player would always say "Yes" as long as they were offered anything. Because otherwise they would be turning down money for nothing, which appears to be a bizarre decision. If you ask people, "Would you turn down money for nothing?" they usually say, "Of course I wouldn't." But in practice, unless the first player offers the second player around 30 percent or more of the prize, they usually do say no. These figures apply in the United States and Europe. The split varies between cultures, but there is always a percentage below which most people would say "No." People are prepared to lose money to punish the other player for being unfair. However, there's a way to turn the game around and show that the psychologist hasn't got a perfect grasp of the situation either.

When I do my talk, after running a conventional ultimatum game I ask the audience to imagine we're playing it again, but instead of it being funded by a psychologist, this time it is being funded by a billionaire, who has put up a stake of $10 million. (In practice I have usually done the experiment with £10 million, but the impact is similar.) Realistically the second person would be very unlikely to turn down an offer of $100,000, for instance, even though the first person would be getting $9,900,000 and the second is only receiving 1 percent. So I get the audience to stand up and start telling them a decreasing amount that they would get as their split of $10 million. I ask them to be honest and sit down at the value at which they would say "No," and turn down the cash.

As there is no real money involved, because I am yet to find a billionaire willing to fund the experiment (offers to @brian clegg on Twitter, please) I think that many people exaggerate how much they would turn down. But typically the response goes like this. A few people sit down above the $50,000 mark. By somewhere in the $10,000 to $5,000 region, around half the audience has sat down. When it's down to $500 a significant majority are seated. And at $1 I have between 1 and 4 stubborn folks still standing. It's a fascinating experiment when you think just how much money people are prepared to give up (or at least say they are) in order to take revenge. But the experiment reflects the same effect as my strange lottery with those alternate expected returns of a 9/10 chance of winning $11.11 or a 1/100,000 chance of winning $1 million. The handful of people left standing at the end of the ultimatum game are often teenagers or children. For them, $1 is worth significantly more than it is to a middle-aged audience member.

This leads us neatly back to the Bernoulli family, and the way that a family member pointed out a flaw in the expected value concept. The mathematician in question was Nicolaus Bernoulli, son of Johann and brother of the most famous member of this overachieving Swiss family, Daniel. Nicolaus considered the outcome of another simple game. All we do to play the game is take note of a series of coin tosses. The amount a player wins depends on what comes up. Each time the result is a tail, the prize doubles and the games goes on. As soon as the result is a head, the game ends with the player winning the value that has accumulated to that point.

So, for instance, if we start with the prize set at \$1, if it's heads on the first throw, you win \$1. If it's tails, you go on to another throw. If heads comes up on the second throw you win \$2. If you make a third toss and the result is heads this time, you get \$4. If you get another tail, then the winnings are \$8 if the next toss is a head . . . and so on. The thing that makes it interesting, Nicolaus pointed out, is to ask how much of a stake you would be prepared to put up to take part in the game? All we should need do is to calculate the expected value and as long as your stake is less than that, it is worth taking part.

To find the expected value we take the probability that the first head will occur on each throw and multiply it by the amount you would win in that case—then we add together all the options. So we have a 1/2 chance of getting a head on the first throw. In that case, the prize will be \$1, so the contribution to the expected value in this case will be $1/2 \times \$1 = \$0.50$. The chance of getting tails on the first throw then a head on the second throw is $1/2 \times 1/2 = 1/4$. And the prize would be \$2. So the contribution to the expected value would be $1/4 \times \$2$ or \$0.50. If we got as far as the third head, the chance would be 1/8 and the prize \$4, giving an expected value of $1/8 \times \$4$ or \$0.50. There's a pattern emerging. The expected value at each stage is \$0.50.

So to get the overall expected value, we simply add together all the contributions from each possible win, which gives us the series:

$(1/2 \times \$1) + (1/4 \times \$2) + (1/8 \times \$4) + (1/16 \times \$8) \ldots$ or to put it another way

$\$0.50 + \$0.50 + \$0.50 + \$0.50 \ldots$

Now, bear in mind that ". . ." means to carry on without

stopping. That should mean that however much money it costs to play this game, as long as you are guided by the expected value, you ought to play. If, for example, it costs $1 million a play, you should do so, because the sum of $0.50 + $0.50 + $0.50 + $0.50 . . . is more than that. It's more than any number. The limit of this series is infinite. The game has an infinite expected value. But the problem that Nicolaus Bernoulli underlined was that the expected value is only really useful if the process is repeated very many times. It can only tell us so much about any particular instance.

It's hard to imagine anyone sensible paying $1 million to play a game where they have a 50 percent chance of winning just $1. What the player needs to do instead is to think what chance of losing they are prepared to take. We know, for instance, that there is a 50 percent (1/2) chance of winning no more than $1, a 75 percent chance of winning no more that $2, an 87.5 percent chance of winning no more than $4, a 93.75 chance of winning no more than $8, and a 96.875 percent chance of winning no more than $16. So to invest even $16 is a long shot.

For fun I just played the game by tossing a coin. The outcome? It came up heads on the third throw—a $4 win. All the tools of probability have their place. At the simple end, the ability to work out the chance of getting a number from either of two dice is handy indeed; for instance, if you are playing backgammon. Expected value is essential in working out whether or not to take on lots of financial commitments. But each tool has to be used with an understanding of its implications for a specific game or investment, not just "averaged over a lot of people" or "averaged over a lot of transactions."

We wouldn't, for instance, be happy with a bank system that was usually good, but every 10,000 transactions would lose all the contents of your account. It really is no reassurance that 99.99 percent of transactions work perfectly if your account happens to be one of the ones that is accidentally emptied. This is why performance statistics that tell us that, say, 99 percent of cases are handled well, are highly dependent on the outcome of them going badly. If it's trivial, like on-time delivery of a burger in a fast-food restaurant, then that statistic is very good. If it's the chance of dying when going into a hospital for a routine checkup, we would be worried indeed.

Where probability-based statistics proved extremely valuable was in dealing with a mass of data, or a mass of participants, whether that mass represented "the American people" or "the molecules of gas in a cylinder." In each case, as long as there was no need to worry about the impact on individuals, the statistical approach enabled mathematics to give us eerily correct predictions of how the mass would behave.

The Scottish physicist James Clerk Maxwell, as we will see in the next chapter, was one of the first to drive much of his approach to science from mathematics. And he was also one of the first to apply statistics to the behavior of gases. The starting point for his work was the matter of smells. Why does the odor of a smelly object (bad or good smell—it doesn't matter) take so long to get from the object to our nose? It was known by the nineteenth century that the molecules in gases fly around at high speeds, traveling at hundreds of meters (or yards) per second, yet a smell typically took a number of seconds to percolate across a room.

The German physicist Rudolf Clausius had suggested that the problem was a matter of collisions. Although molecules were indeed traveling very quickly they were constantly colliding with each other and bouncing off in a new direction. This meant that for a new set of molecules—the "smell molecules"—to diffuse out through the air would take a long time, because they would typically only travel a very small distance before an impact resulted in a change of direction.

Clausius assumed that gas molecules all traveled at the same speed, but this didn't make any sense to Maxwell. It seemed much more likely that the speeds of molecules would vary—some higher, some lower, with their speeds distributed in a curve with a peak somewhere between maximum and minimum. Maxwell realized that the only approach that could be taken, if this were true, was a statistical one, enabling him to get an overall picture of how the molecules behaved. This so-called Maxwell distribution gives a manageable way to make calculations on the speeds of gas molecules dependent on the temperature, transforming the way that the behavior of gases could be predicted.

This ability for statistics to give us an overview of many varying individuals is just as important when dealing with people as it is with molecules. This is what makes it possible to get a picture of what is happening in a large group of people and to predict everything from garment sales to medical requirements. However it is important to be aware of the limitations. Even the statistical behavior of molecules can result in misleading outcomes. If we look, for instance, at the second law of thermodynamics, which says that heat flows from hot to cold, and disorder stays the same or increases in a closed system, it tends to be treated as if it were an unbreakable fact, but in practice, it is statistically based.

The implication of the law is that, for instance, if we open a partition between two boxes of gas, one warmer and one cooler, then over time the two will mix and we will end up with a uniform gas at an intermediate temperature. This is what the second law predicts (the more ordered state of two separate sets of molecules, selected by temperature, giving way to the disordered mix). However in principle there could be, purely randomly, a temperature gradient briefly reestablished. Purely by chance, more hot molecules could head into one box than into the other. It's unlikely to have a major impact with so many molecules involved, but it could happen. Statistics provide us with an overview likelihood, not a certainty.

When dealing with people, the particular danger is not just applying the typical pattern to an unusual grouping, like those gas molecules, but applying the statistics of a mass of people to one individual. We rarely need to consider what happens to a single gas molecule in a whole cloud of gas because they are effectively identical—but people are not. There is a notorious case in the history of statistics when in 1999, British mother Sally Clark was convicted of murdering her two baby sons. She would remain in prison for nearly four years before the verdict was overturned. She was convicted on the basis of a terrible misuse of statistics—both in terms of the mathematical competence involved in the calculation and in the way an incorrect leap was made from the big picture of statistics to the specifics of an individual person's situation.

The trial took place after a second of Clark's baby sons, each aged under three months, had died of Sudden Infant Death Syndrome (SIDS), sometimes referred to as "cot death." A high-profile pediatrician, Professor Sir Roy Meadow, was brought in by the prosecution as an expert witness. Unfortunately, Meadow's expertise did not extend to probability and statistics. A study had put the chance of a child dying of SIDS in a household with no contributory factors as 1/8,543. Meadow told the jury that this meant that the chances of both of Clark's sons dying from SIDS could be discovered by multiplying the probability by itself, making it a 1 in 73 million occurrence, something that Meadow claimed should only happen by chance once every 100 years.

This evidence was central to Clark's conviction. Yet the use of numbers was impossibly flawed. As Cardano had discovered, multiplying together the probabilities of two unconnected events is the correct way to work out the combined probability. So we know that where the chance of throwing a six with a single die is 1/6, the chance of throwing six twice in a row is $1/6 \times 1/6 = 1/36$. The two throws are unconnected. There is nothing about the circumstances of the first throw that has any influence on the second.

However, what failed to come out in the trial is that this use of the mathematics simply doesn't apply in the case of a condition like SIDS. There is strong evidence that this is a condition where two such deaths would not be unconnected events. If a death has occurred, it is far more likely that there would be a second occurrence than simply taking the chances of SIDS striking in the general population. Research published soon after, in response to the trial, suggested that double SIDS cases would occur in the UK not every 100 years, but more like every 18 months.

Leaving aside the bad handling of probability, the other problem with the prosecution case is that it went from "something has a low probability of happening by chance" to "something did not happen by chance." This is a totally unjustified leap that is based on a profound illogicality. Take, for example, a major lottery like the EuroMillions game. The chance of winning the main prize in this is 1 in 116,531,799—it is ridiculously unlikely—yet the game is won most weeks. With a big enough pool of people involved, unlikely things happen time and again. It's not enough to say that because something is unlikely, it is not going to happen.

Apart from the simple fact that with a large enough population involved, something with a low probability will almost certainly happen, the final error was that even if there had been a 1 in 73 million chance of the deaths occurring at random (and bear in mind that 1 in 73 million was overinflated), the inference that was being drawn was that in all the other 72,999,999 in 73 million cases the deaths would be as a result of murder. What the prosecution should have been comparing was the probability of this happening randomly with the chances of a mother in the UK committing child murder twice—which certainly wouldn't have been 72,999,999 in 73 million. Statistics are hugely valuable in science (and the law), but they have to be handled properly.

It is very clear that we can't apply the same kind of predictive statistics we see in the second law of thermodynamics to human behavior, appealing though the idea might be. Even though there are psychological reasons why we tend to get "groupthink" or mob behavior, a collection of people is far more complex in behavior than is a collection of gas molecules. The great American science-fiction writer Isaac Asimov based his Foundation series of stories on a concept he called psychohistory, where an extremely powerful form of statistics is able to predict the development of the culture it studies, down to individual events. But this could never work in practice.

Asimov was inspired by the classic history book *The History of the Decline and Fall of the Roman Empire* by Edward Gibbon. This seemed to show that specific indicators could be used to predict the way that an empire would fall apart. Asimov extended the (already shaky) concept to a whole mathematics of behavior. But in reality, anyone trying to use statistics to predict the future of anything as complex as a civilization comes up against the same problem faced by those who try to make long-term forecasts of the weather. The system is far too complex, with too many variables involved, to make meaningful predictions. It is highly chaotic in the mathematical sense. This means that small changes in starting conditions—typically, in the case of a population of humans, caused by the actions of individuals—have a huge impact in outcomes.

Probability and statistics have become immensely valuable tools in the armory of all scientists. However, this has proved a problem time and again when the scientist in question does not have sufficient expertise in the mathematics and misuses the statistics. There is no doubt here that the math is useful, and does play an important role in the understanding of the science, but there is a danger if too great a weight is put on statistics as a mechanism of "proof," which does not help science and can result in taking an unrealistic leap from the apparently correct numbers to thinking we know more about the reality of the universe than we do.

Some of the problems are not due to the mathematics at all, but rather to the way that they are employed. A lot of these come up in the kind of paranormal research I covered in my book *Extra Sensory*. Say we are running a trial for telepathic ability. We select the people using a preliminary test and work only with those who get high scores. This is fine, as long as we throw away their initial test results. But what commonly happened in trials is that the results from the selection tests were included with the "real" tests. As these results were specifically selected because they were good, they inevitably bias the overall figures toward a positive outcome.

This is a form of "cherry-picking," a common problem where

statistical methods are used. If only data that supports a hypothesis is used and the rest is ignored or given less weight, the outcome is useless—yet it is frequently done. Sometimes this is blatant and intentional. In other cases, like including the selection trials, it can be done accidentally without realizing that the outcome is being biased. Another way this might happen is to look for reasons to discard data. It is fine to discard data because something went wrong during an experiment as long as the data has not been examined. Once it has, though, there is always a danger of finding a reason to drop the data, even subconsciously, as the data didn't support the desired outcome.

Here's another example of unconscious cherry-picking that was produced by an early parapsychology experimenter, J. B. Rhine. In his years of experimentation, Rhine ran many telepathy tests with a series of individuals. These usually consisted of running through a pack of "Zener" cards, each displaying one of five possible symbols, with the subject attempting to transmit the card value to another person by telepathy. These cards came in packs of twenty-five. One test subject, Mr. A. J. Linzmayer managed to guess fifteen cards in a row correct. Rhine announced triumphantly that "The probability of getting 15 straight successes on these cards is $(1/5)^{15}$ which is one over 30 billion."

Here the cherry-picking involved is quite subtle, because Rhine selected that particular run. If he had literally done the test once with the fifteen cards, then the probabilities he cited would technically have been correct, but a single run wasn't enough to establish good reliability. This meant that the highly successful run was just part of one test of many hundreds. The test with the run of fifteen cards wasn't selected at random. It was specifically chosen because it had this sequence. Making that choice was, itself, an act of cherry-picking.

Apart from cherry-picking, there is a wide range of statistical methods, not all of which apply in all cases, yet it is not uncommon for an inappropriate method to be used. Perhaps the most common issue is with small samples and the selection of the sample. Many "soft science" experiments are done with a small number of participants, which means it is rarely possible to make any definitive conclusions from the outcome. The selection part comes in because it is easy to select participants that reinforce a particular view. A blatant example (that has happened) would be for someone taking a survey of what car people liked best to choose a sample of people who currently owned one particular model of car. They inevitably would not represent the population as a whole.

Even the design of an experiment can bias the results. In some areas of psychology and parapsychology testing, the experimenters are looking for very small deviations from the expected norm. Let's say that they are undertaking an experiment where purely randomly you would expect that you would get result A 50 percent of the time and result B 50 percent of the time. And let's run the test multiple times to get a better picture. Say twenty-five times. By choosing an odd number, immediately there is no way to get an exact 50/50 split of outcome.

There are plenty of tests to show how likely the data collected would have occurred by chance. And you will often see in reports that a particular psi ability must surely exist, because the probability of these results occurring by chance are very low. However, in presenting this information, the experimenters make a leap too far. Apart from anything else, psychologists tend to use much lower standards of chance prevention than physicists. It's common for psychology experimenters to consider an outcome that had a 5 percent chance of occurring randomly, to not be random—but, of course, such things happen all the time. Worse, though, just because they have shown that the results were unlikely to be the result of random chance does not show that a particular hypothesis is true—that they were caused by psi abilities.

This aspect of interpretation can result in researchers who have made perfectly acceptable use of statistics to still find it extremely difficult to explain just what their results mean. That was certainly the case in the discovery of the Higgs boson at CERN's Large Hadron Collider (LHC). Finding a Higgs boson, predicted to exist by the standard model of particle physics to give some of the other particles their mass, is not like finding a rare tiger in the wild. You can't catch it, take photographs, or take a blood sample and test its DNA to make sure that it is what it appears to be, as you could with the tiger. Apart from anything else, the experiment does not actually show us a Higgs boson at all. What it does instead is to reveal the indirect traces left by other particles that are assumed to be the result of the Higgs particle decaying. Because of the indirect nature of the discovery, all researchers can do is to offer a probability of what had occurred—and this is where the problems begin.

The way that probability is often presented by scientists is in terms of "sigmas," which refers to the symbol for the statistical measure called standard deviation. If you imagine plotting a distribution of the number of times particular outcomes occur randomly from a particular event, they often form a bell-shaped curve in what is called a normal distribution. Think, for instance of the range of weights of cell phones. They will mostly fall within a fairly tight range and will tend to be distributed pretty evenly around the average weight.

Not every random occurrence obeys a normal distribution,

because the nature of the information being plotted can mean that it forms a different shape. Textbooks sometimes give human height as an example of a set of data that fits a normal distribution. But it doesn't. The average height of a U.S. male is around 5 foot 6 inches, which gives a clue to what's happening here, because the *typical* U.S. male (the median height in statistical speak) is taller than this. The right-hand side of the distribution, where height is getting bigger, only extends for about a foot above the average before the numbers get tiny. There are very few men taller than 6 foot 6 inches. But the left-hand side of the distribution extends more than 2 feet below the average. The outcome is that the distribution is not truly normal, but is "skewed" to the right with a long shallow tail to the left.

Standard deviation is a measure of the shape of a distribution (it is only a helpful measure in a symmetrical distribution like the normal distribution). The standard deviation tells us how the distribution of occurrences spreads out—whether there is a very wide spread of data, or whether most are close to the mean (average). If a particular outcome is given a 1-sigma level, then just over 68 percent of the time we would expect the outcome to fall within that range as a random occurrence. With a 2-sigma level it would fall within that range around 95 percent of the time. This is the kind of level often adopted in the "soft sciences" like psychology. However, for the Higgs boson, the 5-sigma level was adopted. In that case we're looking at an occurrence that would only fall outside that range in 1 in 3.5 million cases. But how can we describe that in terms of the certainty (or otherwise) of actually having discovering the Higgs boson?

This caused a terrible minefield for the media when they tried to explain what had happened. What the statistics show is that it is very, very unlikely that the results produced by the LHC were produced by chance alone. But like the Sally Clark case, we can't just invert this to say that the unlikeliness of these results occurring at random means it's very, very likely that the Higgs exists. It doesn't actually show it's likely that a Higgs exists, merely that there was very likely a cause, which we are assuming to be a Higgs.

To make matters even worse there is a very subtle distinction here that it's almost impossible to avoid tripping over. Some news outlets reported that the results meant that there was a 1 in 3.5 million chance of there not being a Higgs boson. But in reality the statistics show there was a 1 in 3.5 million chance of this data occurring without a cause. The measure doesn't say how unlikely it is that this outcome happened as a result of chance. It says how unlikely this data is, if there were no cause. It's the difference between saying "given these results, the chance there is no cause is very small" (wrong) and "given a totally chance occurrence, these results would only occur very infrequently" (right). That subtle difference in stress makes a significant difference in terms of its value to science.

What seems to be the case with probability and statistics is that, applied correctly, they can have a good match to reality. This makes sense. We are not truly modeling a real-world process in abstract mathematics here. Instead we take a numerically based fact, or near-fact about the physical world—for example "a fair coin has a 1/2 probability of tossing a head and a 1/2 probability of tossing a tail"—and use the numerical methods that are relevant when this numerical fact is the case. We are not so much applying mathematics to the universe as applying mathematics to numbers.

When probability and statistics are applied correctly, the problem tends to be that understanding does not come naturally to us. Because we recognize and understand the world through patterns (see page 28), we see patterns where they don't exist all the time. This makes us uncomfortable with the real nature of randomness and distributions of events, making probability-based statistics a tool that has to be handled with care, even by the professionals.

Probability and statistics have proved increasingly important to mathematically based physics—and not just in the hunt for the Higgs. But before probability proved to be at the heart of the nature of the tiny particles that make up all matter, another tipping point was reached in the way that mathematics was driving scientific thinking. A tipping point that had light at its heart.

11 Maxwell's Mathematical Hammer

A few years ago, I took part in a debate at London's Royal Institution (RI). The idea was, in a fun way, to decide who was the first scientist. It is no coincidence that all four of the characters put forward in that debate have a place in this book. In the debate, the early work of Archimedes and Roger Bacon (my candidate) was trumped by Galileo, whose modern scientific credentials made him the winner. But there was one much later candidate, James Clcrk Maxwell, put forward by the resident historian of science at the RI.

On a trivial argument Maxwell could take the title, as the word "scientist" was not coined until his time. Up until then, the accepted term was the clumsy "natural philosopher," but it was argued in 1834 that just as an artist worked in art, it seemed reasonable that a scientist should be a person who worked in science. (Thankfully they didn't go for "savant," which was one of the alternatives suggested at the time.) But the argument presented on Maxwell's behalf in the debate was more subtle than this. Maxwell was the first in the field of science who put forward a theory where the mathematics operated in true abstraction from reality.

Maxwell was not the first scientist to use mathematics of course. Newton, as we have seen, wove magnificent scientific spells with his math. But Maxwell's work on electromagnetism, which also defined the nature of light, was more tightly tied to the mathematics. The final version of his stark, beautiful equations stood aside from any consideration of the physical reality, derived from a pure mathematical formulation. And so began a painful separation. Newton's basic mathematics is surprisingly simple if you can get past his deliberate attempts to make it obscure. But Maxwell's work reaches the state where his results are incomprehensible to the casual observer, who has to then take the theory's predictions on trust, which has important implications for both science and the support of science in society.

If you haven't heard of Maxwell, it's not particularly surprising. It would be interesting to take a poll of physicists, asking them for three names: the greatest physicist in history, their favorite physicist of all time, and the physicist most underrated by the general public. Isaac Newton or Albert Einstein would be likely to win the first category. Richard Feynman would top the second poll hands down. But James Clerk Maxwell would be most likely to triumph as that underrated individual. It's notable that Einstein had three pictures on his wall: Newton, Faraday, and Maxwell.

Maxwell's life story has been told often enough in popular science books, though it is worth noting where his unusual name came from. Maxwell's father John was born John Clerk, but John's father inherited an estate and title from the Maxwell side of their family. When John's father died, the title and the main Middlebie estate went to John's older brother, while John took on the subsidiary Glenlair estate, at which point he tacked "Maxwell" onto his name to emphasize the connection, so James was born a Clerk Maxwell, but the name is not hyphenated and is shortened to Maxwell.

After a privileged upbringing that allowed much freedom to go his own way and explore the natural history of the countryside around his home, Maxwell began a physics degree at Edinburgh University, but transferred to the more prestigious Cambridge, where he began a career that would take him up to Aberdeen and down to King's College, London, before spending a period of time financing his own work on the Glenlair estate. He apparently enjoyed the lack of pressure that came from working outside of a university, and it was only the opportunity to become the first Cavendish professor at Cambridge, designing a brand-new laboratory that would push Cambridge to the heart of the physics of the day, that drew him back into formal academia.

Along the way, Maxwell worked on a whole range of topics, notably on the use of statistics in thermodynamics (see page 133), the use of polarized light to monitor strain in transparent materials, and the nature of color perception, even producing the first ever color photograph. But his real legacy was his work on electromagnetism. Maxwell's predecessor as the master of electricity and magnetism, Michael Faraday, had come up with a revolutionary way of considering these phenomena by imagining that electrical charges and magnets produced a field. His concept of a field was something like a contour map, filling all space, where the values of the contours reflected the strength of the electromagnetic phenomena. When a wire cut through the contour lines of a magnetic field, it would produce an electrical current because of its interaction with the field.

Although Faraday was without doubt a visionary scientist, he had no mathematical training and made very little use of math in his work. Other scientists had applied numbers to electrical and magnetic effects, but they treated them as an indirect "force at a distance," just as Newton had considered gravity. Maxwell was arguably the first to see the real power of Faraday's field concept, and found a way to apply mathematics to the concept of fields that would transform the understanding of electricity and magnetism.

To model a field mathematically required a way of adding up its influence from many points in space, because the extravagant thing about fields was that they extended infinitely through three-dimensional space. Of itself, the mathematics was straightforward stuff—the calculus of Newton and Leibniz was designed for exactly this kind of operation (though rarely in three dimensions). But Maxwell had to go one step further, extending the reach of calculus. In Faraday's "force fields" like those of magnetism and electricity, the value at each point in space was the force produced by a magnet or electrical charge. And a force is a vector. Unlike, say, mass, which is just a number (referred to as a scalar), a force has both a size and a direction. Mass just sits there and is, but a force has to act in a particular direction. A vector is a double mathematical value that combines two quantities. In the case of a force, it describes how strong the force is and in which direction it pushes.

At the time Maxwell was working, the mathematics of vectors was just starting to be developed—a necessary extension because it was pretty obvious that a vector couldn't just be treated like a simple number. It was known, for instance, that to add two vectors was more like an exercise in geometry than arithmetic. The simplest way to add vectors was to draw two arrows, with the length of each arrow representing the size of the vector (the force, or instance) and the direction of the arrow indicating the direction in which the vector is pointing. The process begins by drawing an arrow representing the first vector, then drawing the next arrow starting at the head of the previous one. The result of the addition was found by drawing a third arrow replacing the other two that ran from the starting point of the first arrow to the head of the second. However, Maxwell needed more than simple addition of two fixed values. To explore the calculus of vectors he had to deal with changing values that varied from point to point throughout space, and it was by using this that he began to work on electromagnetism.

Initially Maxwell did not move too far from the mechanical pictures of the world that were used to model pretty well everything at the time. He played with describing electromagnetism (the combination of electricity and magnetism) as a fluid flowing through an imaginary porous solid, with the flow of the fluid representing the lines of electromagnetic force. Like all good models, his attempt not only matched the observed values that were used to produce the model in the first place, but also made predictions that could then be verified, notably predicting the way that the strength of the fields fell with the inverse square of the distance from their source.

His "fluids in a porous solid" model was a useful first tool, but only went so far. Most significantly, the fluid flows, representing the lines of force, were fixed in position by the channels through the solid. But in the vast majority of cases, for example in Faraday's increasingly successful electric motor and generator, the lines of force were moving. To model this would need a total transformation of Maxwell's model, which as yet he could not fathom. It would be a good five years, in part because he packed up and moved down to London at around this time, before he could rework his ideas with any success.

A lesser scientist may well have tried to modify his existing model to deal with moving fields, but it's often the case that great scientists are those who realize when to throw away their old approach, even if they had invested considerable time and effort into it, and start again with something fresh. In this case, Maxwell moved from a model involving the movement of fluids, even today among the trickiest aspects of physics to deal with accurately, to the better-understood field of mechanics. He began by looking at magnets and the need for his model to produce magnetic lines of force that would have tension in one direction, corresponding to the attraction of opposite magnetic poles, and pressure at right angles, corresponding to repulsion of like poles.

He imagined that a magnet was made up of a collection of tiny cells that were free to rotate. When a real body, like the Earth, rotates, it bulges around the equator and squashes up a bit at the poles because of the rotational forces acting on it. This would also happen to his tiny cells. If a collection of the cells had the same axis of rotation, then there would be tension along that axis as the cells squashed up, and outward pressure at right angles where the cells were bulging, exactly the effect required to model the action of the magnet.

So far, so good, but in practice the cells would tend to seize up as they interacted, and Maxwell was missing the electrical part of the equation. To allow for this, he envisaged that there were far tinier spheres surrounding each cell, in the same way that ball bearings are used to prevent friction around rotating axles. Unlike the cells, which could only rotate in place, these little balls could flow through the material, corresponding to the flow of electrical current. Although it was only a crude representation, it was startling how close Maxwell came to a picture of metal composed of atoms with electrons flowing through it, well before the electron, or even evidence for the existence of atoms, was discovered.

At this stage, Maxwell's model worked well for some aspects of electromagnetism, but couldn't explain induction, the essential physics behind the transformer—how a changing current in one wire produces a surge of current in another. Faraday had correctly deduced that switching the current on or off produced a magnetic field, which expanded out and contracted back through the second wire, producing electricity. Maxwell managed to use his cells (which by now he had transformed to be hexagonal) and tiny balls to model this operation. By considering the way different layers of cells would interact as the balls flowed around them and by adding in a resistance to the balls' movement that meant they would slow down over time, he extended his model to produce induction.

With this addition, the only aspect of electricity and magnetism that his model did not adequately cover was the interaction between electrical charges. This is where, for instance, a force is generated between small pieces of paper and a comb, which has been electrically charged by running it through dry hair. Like many others before him, Maxwell found that taking a little time away from the problem brought the solution to him with relatively little effort.

He had assumed that in an insulator like paper or ceramic, the little balls were fixed to the cells, so the balls could not flow as they did in a metal. But if he imagined that the cells in an insulator were rubbery—that they could twist a little in place—

then the twist in the cells could act like a spring, storing up energy to be released. By comparison, the cells in a metal would be rigid, with very little tendency to twist. Not only did this work and was able to model the behavior of real materials with startling accuracy, it also gave a real insight into the nature of electromagnetism, with the electrostatic force being more like the potential energy of a spring, while the magnetic force was more like rotational energy. And neither could act in isolation. There was always an interplay between the two. All was going extremely well, but then Maxwell noticed something highly unlikely that could have made his whole model worthless.

Faraday's fields were assumed to be everywhere—even in empty space. And since space was a good insulator, it ought to have elastic cells according to his model. One of the characteristics of elastic things is that you can send waves through them—in fact waves require elasticity to propagate. So this seemed to imply that it would be possible to send electromagnetic waves through empty space. What's more, the twist of the elastic cells would produce a magnetic field, twitching the adjacent little spheres to produce an electrical field. There should be a self-sustaining wave with magnetic and electrical components at right angles to each other and the direction of travel. An electric wave would produce a magnetic wave, which produced an electric wave . . . and so on indefinitely.

It had been established by then that light appeared to be a wave that, uniquely, wiggled from side to side in the medium it moved through. (Usually a wave had to be on the edge of a medium to go side to side.) And it was known from experiments that there seemed to be a relationship between light and magnetism. Was it so unreasonable, then, Maxwell thought, that light *was* his electromagnetic wave? After all, light effortlessly crossed the empty space from the Sun to the Earth. As an impressive reinforcement of his model's power, Maxwell estimated the speed of his hypothetical waves and found that it was near enough the same as the speed of light in a vacuum.

There was no doubt that Maxwell's achievements to date had involved an impressive use of mathematics—far beyond the reach of most of his predecessors, but after a brief dalliance with the viscosity of gases, he came back to have a further attempt at electromagnetism, throwing away his model once more and starting from scratch. And this is where he made the leap to an approach that would be very familiar to the modern physicist by attempting to build his model not from an analogy, like the flow of a fluid or the rotation of mechanical cells, but from pure, unadulterated mathematics.

There was still a model, but in this case the model was simply a set of numbers that behaved according to a collection of logical rules. There was no picture, no analogy; there was no easy mechanism to grasp what was going on. Maxwell brought to play a method devised in the previous century by the Italian mathematician Joseph-Louis Lagrange. This so-called Lagrangian (a contraction of Lagrangian function) made it possible to write a series of differential equations (linked equations making use of calculus) that described how a system that changed with time evolved, based on factors like the momentum of its components and the kinetic energy in the system.

The elegant (to a mathematician) thing about the Lagrangian was that it was a black box. To use it was simply a matter of putting in the known factors, turning the handle and out came results—without ever needing to know any details about the physical nature of the system. It was a modeling process based solely on numbers.

The result, after some complex manipulation and the need to extend Lagrange's work for the special requirements of electromagnetism, was a set of relatively simple equations that between them described the way that electricity and magnetism behaved. Maxwell's original version of these equations was refined and compacted by his predecessors, using more modern notation to produce these frightening-looking four, short equations:

$$\nabla \cdot \mathrm{D} = \rho_f$$

$$\nabla \cdot \mathrm{B} = 0$$

$$\nabla \times \mathrm{E} = -\frac{\partial \mathrm{B}}{\partial t}$$

$$\nabla \times \mathrm{H} = \mathrm{J}_f + \frac{\partial \mathrm{D}}{\partial t}$$

To a modern physicist these equations are straightforward, bread-and-butter stuff (though most would admire their sophistication). But at the time he produced his original version, Maxwell's ability to model with mathematics alone was considered daunting by most scientists. William Thomson, Lord Kelvin, a significant contemporary of Maxwell's who had become a physics professor at an even younger age, was among many who struggled to deal with this degree of abstraction. It didn't help that no one had seen electromagnetic waves other than light, yet the theory suggested that such waves should be generated simply by sending an electrical charge up and down a piece of metal (an aerial as we now call it). It was a good twenty years before Heinrich Hertz produced the first radio waves this way, helping to make Maxwell's remarkable achievement a solid success.

Although mathematics had made Maxwell's work possible, at this stage even he was not prepared to take on all the implications that came out of the formulas. He might have modeled with mathematics, but he wasn't assuming that the numbers were a direct basis for the nature of reality. In two separate ways, Maxwell was prepared to turn a blind eye to what the numbers said. One was in making an assumption—the other in simply ignoring one of the predictions that came out of the equations, because it seemed too strange.

The assumption was about the ether. Maxwell's electromagnetic waves depended on empty space being able to sustain electrical and magnetic fields—and it was those fields that supported the waves. There was no need for a medium for the waves to travel through. In fact, as mentioned above, had there been such a medium that behaved like an ordinary physical material, the waves would be odd indeed, because transverse waves that wiggle from side to side can't travel through the middle of a medium, as the medium would damp down the side-to-side oscillation. Usually such waves travel on an edge—say on the surface of water, or in the side-to-side vibration of a violin string. When waves travel through a medium, like sound through air, they tend to be compression waves, wiggling in the direction of travel.

However, despite his model clearly predicting that there was no need for the ether so that light could travel through empty space from the stars, Maxwell resolutely stuck to his assumption that the ether must exist. All good scientists are a mix of iconoclast and traditionalist. They need to bring fresh new ideas, replacing old ones, but at the same time they can't start from scratch with everything. They need to build on existing concepts. And all too often, those concepts last long past their sell-by date, as happened with the ether. Interestingly some modern physicists, like the Nobel Prize winner Frank Wilczek, think that in a sense the ether does still exist—as long as we are prepared to consider the existence of the various fields, like the electromagnetic field, that mathematically fill space as a new way of looking at the concept of an ether.

The prediction that Maxwell (and all his contemporaries and successors for well over fifty years) ignored was far more startling than the existence or nonexistence of the ether. It was a requirement for there to be waves that could travel backward in time. To illustrate what happened it is worth briefly thinking about a trivial math problem. What is the value of x here?

$$x^2 = 4$$

Even if the very word "algebra" fills you with fear and loathing, dealing with this equation isn't too scary a proposition. We're just looking for the value of x that, when multiplied by itself makes 4. And it's not too difficult to spot that an answer to this is 2. But if you were to give this as the solution to the equation in a test at school, you would only get half marks, because there is another solution. It is equally valid to say that x is −2. The equation has two solutions, 2 and −2.

This is always the case with some classes of equations, like the familiar quadratic equation, of which this is a simple form. And, as it happened, it was also true of the way that Maxwell's equations predicted the potential for a self-supporting electromagnetic wave to exist. The equations had not one, but two solutions, which were given the names "retarded" and "advanced" waves. According to the equations, when a familiar electromagnetic wave—light in all its forms from radio through to X-rays and gamma rays—travels from A to B, this is the retarded wave. But the equations also describe a second wave, the advanced wave, which travels from B to A, setting off at the moment in time that the retarded wave arrives at B and traveling backward in time to arrive at A just as the retarded wave departs.

There were clearly two huge problems with this. No one had ever seen these advanced waves, and if they did exist, they seemed to perform the impossible feat of traveling backward in time. Although there was nothing about the mathematics that said one solution should be ignored and a particular one should be preferred, this was precisely what everyone did, because the alternative seemed too strange to contemplate. Mathematics had made a prediction that the real world seemed unable to match—yet the equations were so good for the rest of the behavior of electricity and magnetism that they had to be used.

It was only in the 1940s that two American physicists, John Wheeler and Richard Feynman, realized that advanced waves were not only predicted by the equations, but could be useful for physics. Although science requires a degree of open-mindedness to progress, most scientists are fairly blinkered by the current scientific theories. Wheeler and Feynman were both particularly good at ignoring the accepted wisdom.

The time-traveling advanced waves would come in useful in dealing with one of the problems facing quantum electrodynamics (QED), the model that Feynman and others had developed to explain the interaction of light and matter. This approach had a downside in that it tended to produce infinities in the math (see next chapter). This continues to be the case both for QED and its more modern siblings like quantum chromodynamics. An example of the type of problem that was facing QED, when Wheeler and Feynman came up with their audacious idea, was electron recoil. When an electron dropped in energy within an atom and released a photon, the electron recoiled like a gun firing a bullet (photons don't have mass, but they do have momentum, which has to be conserved).

To cause that recoil, the electric field of the electron had to act on the electron itself—and this was exactly the kind of result that tended to shoot off to infinity in what was effectively a feedback loop. Yet it was known that electrons gave off photons all the time—that's where the vast majority of the light we experience comes from. Wheeler and Feynman realized that by changing the viewpoint from the usual idea of a single photon being produced, to thinking of a pair of photons, one being the advanced photon traveling backward in time, the recoil could be explained without producing embarrassing infinities.

There is no doubt that Wheeler and Feynman's approach produced a useful result, but generally speaking this has been interpreted as an effective mathematical trick, rather than a result that has any implications for the nature of reality. Certainly most of those who have made use of it (if not Wheeler and Feynman) have pretty much dismissed the idea that advanced waves or advanced photons truly exist in the physical universe, but it was one more example where the numbers produced a remarkable (and in this case unexpected) parallel to what was actually observed. Our common sense tells us that waves can't travel backward in time, but the mathematics predicts this, and that mathematical weirdness proved more valuable in reflecting reality than a more straightforward approach. Arguably, here, numbers were *more* real than intuition would allow.

To a modern mathematician, whichever way you interpret the solutions, the symbols of Maxwell's equations are relatively trivial pieces of work, even if coming up with them in the first place was a work of genius, but even the mathematical professionals had their limits, as Maxwell's contemporary Georg Cantor would discover.

12 Infinity and Beyond

Not entirely surprisingly, infinity is a topic that never fails to stimulate the mind. Thoughts about the nature and existence of infinity go back all the way to the Ancient Greeks. They were certainly aware that a sequence of numbers like the positive integers, the simple counting numbers would go on forever. If there were a biggest integer—call it max—then there surely could always be max + 1, max + 2, and so on. But the whole idea of infinity made the Greeks uncomfortable. Their word for it, *apeiron*, suggested chaos and disorder.

The Greek philosopher who took the definitive approach to infinity for the period (a point of view that would remain dominant for centuries to come) was Aristotle, born in 384 BC in northern Greece. Aristotle argued that infinity was both necessary and impossible. He used examples of aspects of the universe that he considered infinite. The integers, as we have seen, or the span of time—which he argued had no end. And he believed that something could be divided up an infinite set of times. But equally he came up with a range of often confusing arguments as to why infinity could not exist in the real world. For example, he pointed out that a body is defined by its boundaries. If a body were infinite it would have no boundaries, hence it could not exist.

After what was clearly a considerable mental struggle, Aristotle finally decided that infinity was a potential, rather than a concept that was fulfilled in reality. This "potential infinity" was something that could be aimed for, but could never practically be achieved. Infinity existed, but could not be made real on demand. To illustrate the concept he used the neat example of the Olympic games. The games existed—there was no doubt of that. It wasn't a fictional concept. But generally speaking, if someone asked you to show him or her the Olympic games, you couldn't. The games were a potential entity, rather than something you could point at and identify. Aristotle was careful to point out, though, that some potential entities were going to become actual at a point in space or time, yet this wasn't the case with infinity.

This neutered concept of potential infinity was exactly what Newton and Leibniz (see chapter 9) were dealing with when they devised calculus. The infinity of calculus is something that we head toward—it is a limit that is never practically reached. And the target is exactly what the familiar symbol for infinity, the lemniscate (∞) represents. It is the symbol for Aristotle's potential infinity. The lemniscate was introduced by Newton's contemporary, John Wallis, who had written a rather dull treatise on the three-dimensional shapes known as conic sections, which are the result of cutting a pair of cones positioned point to point along various planes. (No one can accuse mathematicians of not knowing how to have fun.) Wallis just throws in a line that says "let ∞ represent infinity" without ever explaining where this symbol comes from.

For the vast majority of mathematicians, with one notable exception, this was sufficient to carry all the way through to the nineteenth century. In fact, potential infinity was generally considered to be the only respectable way to think about the infinite. For example, Carl Friedrich Gauss, the eminent nineteenth-century German mathematician definitively remarked:

> I protest against the use of an infinite quantity as an actual entity; this is never allowed in mathematics. The infinite is only a manner of speaking, in which one properly speaks of limits to which certain ratios can come as near as desired, while others are permitted to increase without bound.

The exception to this blinkered thinking was the remarkable Galileo Galilei. The first thing that springs to mind when Galileo is mentioned was his championing of the Copernican theory that put the Sun rather than the Earth at the center of the universe, leading to his trial by the Inquisition and permanent house arrest. However, in scientific terms his most significant work was the book he published in 1638 called *Discorsi e Dimostrazioni Matematiche Intorno a Due Nuove Scienze* (Discourses and Mathematical Demonstrations Concerning Two New Sciences). This was his masterpiece of physics, laying the ground for Newton's triumphant completion of this work on mechanics, forces, and movement.

Like his book on Copernican theory that got him into so much trouble, this new work was structured as a conversation between three characters, a format that was very popular at the time. Written in conversational Italian rather than stuffy Latin, it remains far more readable today than the formal and often near-impenetrable work of Newton. Given his position, serving a life sentence for the publication, it was remarkable that Galileo got the book published at all. He attempted to do so originally in Venice, then proud of its independence from Rome, but there was still a requirement to get the go-ahead from the Inquisition, which had issued a blanket prohibition on printing anything that Galileo wrote.

If there was one thing that Galileo excelled in, it was stubbornness. Despite the prohibition, despite the risks of even indirectly evading it, when the Dutch publisher Lodewijk Elzevir visited Italy in 1636, Galileo managed to get a copy of his new manuscript to him. One fascinating aspect of the book as it finally came to print is the dedication. In earlier years, Galileo had always attempted to dedicate his writing to a power figure, who might as a result give him patronage. This book he dedicated to a former pupil who was now the French ambassador to Rome, Count François de Noailles. However, where previously Galileo could simply lavish as much praise as was possible (and plenty was possible in the sycophantic style of the time), here he had to be more careful, as the last thing he wanted to do was get Noailles into trouble with the Inquisition.

In the wording, Galileo combined deviousness with an apparent naïveté. It is highly unlikely that the Inquisition fell for his attempt at deception—though, in practice, they seemed to have turned a blind eye. According to Galileo:

> I had decided not to publish any more of my work. And yet in order to save it from complete oblivion, it seemed wise to leave a manuscript copy in some place where it would be available at least to those who follow intelligently the subjects which I have

> treated. Accordingly I chose first to place my work in your Lordship's hands . . .

So, on the one hand Galileo was thanking Noailles for his help. But at the same time he didn't want to make it sound as if Noailles had been directly responsible for the publication, so he threw in some mysterious intermediaries:

> I was notified by the Elzevirs that they had these works of mine in press and that I ought to decide upon a dedication and send them a reply at once. This sudden unexpected news led me to think that the eagerness of your Lordship to revive and spread my name by passing these works on to various friends was the real cause of their falling into the hands of printers who, because they had already published other works of mine, now wished to honor me with a beautiful and ornate edition of this work.

He could thank Noailles, but also managed to blame unnamed friends of the ambassador for passing the manuscript to the printer. It's clear that the idea that all this had happened without Galileo's knowledge until the book was almost ready to print was a fiction. Not only did he ensure that Elzevir received a copy of the manuscript on his Italian visit, there was a considerable correspondence between Galileo and Elzevir over the content of the book. Galileo was the kind of author that cause publishers to tear their hair out, wanting to tweak his output to the last possible moment before going to print. This is bad enough with today's electronic printing, but was a nightmare when each page had to be carefully set up in movable type and made into a physical printing plate. But whether the Inquisition was fooled or simply looked the other way, it did not intervene and the book was published, if unavailable for sale in Galileo's native Italy.

The "two new sciences" in the book's title were those of the nature of solid matter and an analysis of motion, and it was in the first section that the topic of infinity came up. In trying to understand why solid matter sticks together so effectively—why, for instance, a piece of metal is so hard to break up—one of Galileo's protagonists suggested that it is the vacuum between the tiny particles of matter that held them together. (He was wrong, it is electromagnetism, but it wasn't a bad idea.) This theory was queried by Simplicio, whose role in the book was to challenge new thinking, mostly sticking to Ancient Greek ideas. Simplicio argued that there could only be a tiny bit of vacuum in so small a space, which could only apply a tiny force—far smaller than the powerful force that holds a piece of metal together.

This led on, with the trio thinking about how a very large number of tiny forces could add up to a massive amount. "Thus," said Sagredo, one of the characters, "a vast number of ants might carry ashore a ship laden with grain." Any resistance, he suggested, as long as it is not infinite, could be overcome by a multitude of minute forces. Salviati, who primarily spoke with Galileo's voice, mocked the proviso "as long as it is not infinite," saying that it was perfectly possible to have an infinite number of vacua in a finite object.

This seems to have been an excuse to play with some entertaining ideas on infinity, as Galileo next spent a considerable amount of time exploring the nature of the subject. And this was not Aristotle's feeble potential infinity that kept the mathematicians happy for so long, but the real, naked thing. Galileo illustrated the way infinity works in a mind-boggling fashion using an illustration of a strange imaginary device.

He dreamed up a pair of hexagons, one smaller than the other, with the small hexagon stuck onto the front of the big one, both aligned the same way. Each rested on a horizontal rail. Now, the others were asked by Salviati to imagine rotating the hexagons through 1/6th of a turn. The wheels would move forward by the length of the side of the large hexagon as it shifted onto its next side. This was not surprising. But the small hexagon's sides were much shorter. Even though it had only gone through 1/6th of a turn along its rail, so should have moved forward the length of one of its small sides, it had actually moved forward by the size of the side of the big hexagon.

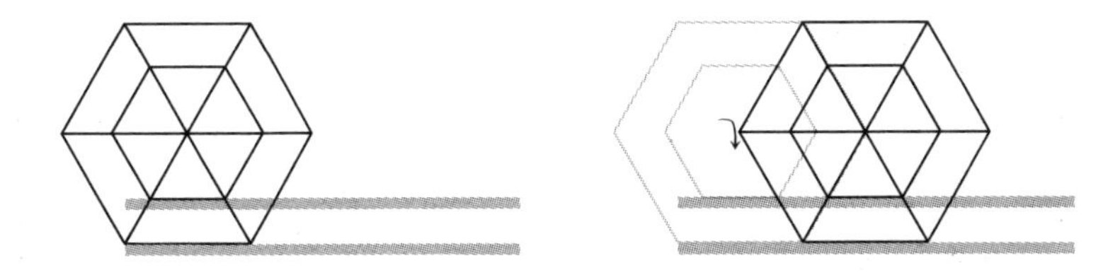

This was possible because as the big hexagon turned, it lifted the small hexagon off its rail and shifted it forward by a jump equal to the difference between the lengths of the two sides. So the small hexagon had both the movement of the length of its side plus the jump, adding up to the length of the large hexagon's side.

So far, so good. Galileo now imagined increasing the number of sides on the wheels, more and more. As he did so, to produce 1/6th of a turn required the new device to turn through more and more of the sides of the large wheel, with more, but shorter, jumps between the sides of the smaller wheel. Here's the clever bit. He imagined going all the way to a pair of circular wheels. In effect, the number of sides has become infinite. If he now turned the wheel 1/6th of a turn, then the smaller wheel had also rotated by 1/6th of its circumference—yet it had still managed to move forward as far as the big wheel. And in this case, because the wheel was perfectly circular, it never appeared to lift off the rail.

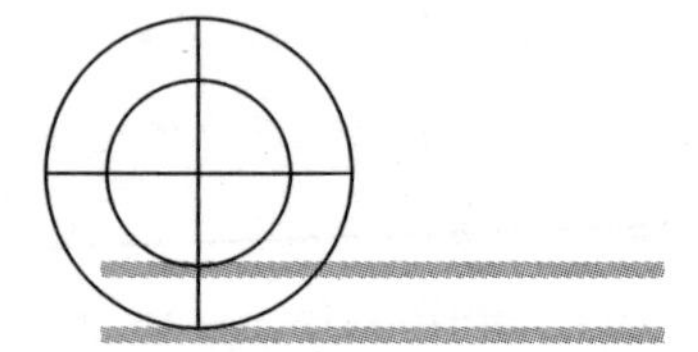

This seemed puzzling (particularly to Simplicio). Galileo in the form of Salviati argued that this is because the small wheel had undergone an infinite number of infinitely small jumps, which collectively added up to provide the extra distance that the small wheel moved along its rail. Salviati, admitted ruefully that this was startling, but then asked permission to have a little deviation from the topic of the book to consider infinity further, and the others were delighted to have the diversion.

After an obscure example involving a geometrical proof that a point and the circumference of a circle could be the same size, the discussion returned to the wheels. Simplicio has noticed that what the first example seemed to be saying was that there were two infinities—the infinity of points around 1/6th of the circumference of the big wheel and the infinity of points around 1/6th of the circumference of the small wheel. Both were infinite in number, yet one somehow managed to produce a result that was bigger than the other. Salviati initially fobbed this off as a problem of trying to understand the infinite with a finite mind, but then set out to prove to Simplicio that this kind of oddity was an inherent part of the nature of infinity.

The proof involved the squares of positive integers—the counting numbers. Every counting number, Salviati pointed out, had a square. Simplicio was happy with this. So Simplicio asked him to imagine the infinite list of such numbers, each of which has a square corresponding to it—so there were clearly the same number of squares as positive integers. And yet, at the same time, there were plenty of positive integers that weren't themselves squares. Numbers like 2, 3, 5, 6, 7, and so on. So there was a square for each counting number, and yet there were far more counting numbers than squares.

What Galileo had realized, and made explicit in his discussion, was that when dealing with the "real" infinity, the conventional rules of arithmetic do not apply. Concepts like "equal," "smaller," and "bigger" lose their traditional meanings. We would now say that an infinite set (the positive integers) can contain an infinite subset (the squares). One reason that Galileo's characters were struggling was that they were trying to treat infinity as if it were a number—Galileo refers to it as such. We now don't think of infinity as a number. We can refer to an infinite *set* of things, but not an infinite number, for reasons that would become clear a couple of centuries later.

After Galileo, all eyes returned to the less scary potential infinity until in the nineteenth century, when Georg Cantor was prepared to take on the real thing. Cantor was a mathematician whose career, and eventually whose mind, was destroyed by a combination of his belief in the reality of infinity and by the opposition of other mathematicians who felt that he was playing with fire. Cantor believed that mathematics and mathematicians could take on the stark, unshielded reality of infinity. His work on the topic, proving the apparent impossibility that there has to be something bigger than infinity, had no obvious practical applications in the real world, but there was more to Cantor's achievements.

He also codified a kind of metamathematics, set theory, that seemed to explain the workings of mathematics itself, and we need to get a feeling for what Cantor did here to be able to appreciate his wizardry with the infinite. Right back in the first chapter we came up against the problem of what numbers are and how they relate to the world around us. Set theory gives a formal definition of the numbers, apparently based on reality and yet capable of being abstracted to stand alone in the mathematic universe outside our Platonic cave. Set theory is roughly the equivalent in mathematics of atomic theory in science. Just as we got along without acknowledging the existence of atoms for millennia, but once we accepted they existed they became the building blocks of our understanding of nature, so we managed to do mathematics quite happily for millennia without consciously using set theory, but once it was developed it became the foundation of the rest.

A set is nothing more or less than a collection of things, whether physical objects or concepts. They can be things that share a common identity—the set of things called "Brian" or the set of things that look like a donut—or a random collection linked only by location or time. The set of things that are on the sidewalk in New York, or the set of things you thought of this morning. Some of the language of set theory has escaped into general usage. A "subset" is a part of a set where the members also share a different common linkage—it is a set within a larger set. So, for instance, the set "Americans" form a subset of the set "Human beings." Each individual item in a set is called a member of that set—so unless you are an artificial intelligence, you form a member of the set "Human beings."

You've probably seen visual representations of sets in the form of Venn diagrams. They can be handy to understand the way that sets intersect and combine. So, for instance you could have a diagram showing the intersection between the sets "Human beings" and "Things that live in New York," many of which will not be human. The overlapping segment represents "Human beings who live in New York."

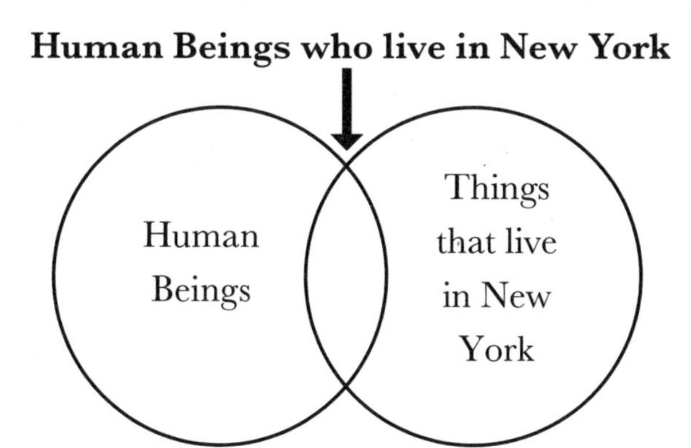

We often unconsciously handle sets when we use a search engine. When we use terms like "AND" and "OR" and "NOT" these so-called Boolean algebra terms are mechanisms for combining or selecting from sets. So if you were searching online for images and put in the following search terms:

(car AND American) (Ford OR Chevrolet) (NOT red)

we would be selecting the subset of "cars" that was American, either Ford or Chevrolet, and any color but red. At least, that used to be the case. These days a search engine like Google or

Bing considers itself too clever to be told what to do, so they rarely follows these Boolean terms literally.

When dealing with sets, there is a distinction between two possible ways that we can use numbers that need to be clarified. For most of this book we have used numbers like 1, 2, and 3 as counting numbers. These are the "cardinal numbers" meaning that the numbers are being used in their principal role. However, it is also possible to use numbers to specify the position in an order that a member of a set holds—which are the "ordinal numbers." So when considering a set of oranges, for instance, the number 3 could either refer to the number of oranges in the set, or the third orange in the set ("orange number 3").

We tend to think of ordinals as relatively minor though valuable uses of number, but some anthropologists have suggested that ordinals came into use before cardinals. This would require a very different picture to our goat counting in chapter 2. The suggestion is that counting was not first used for something so mundane as trade, but rather for religious rituals, where the important thing was to get stages of a ritual in the right order, and so it was the numbering of order that came before the numbering of objects. There is no good evidence for this, and it could easily be seen as anthropologists trying to claim greater significance for their outlook on life over that of bean counters—but it certainly is likely that ordinal counting came in relatively early, even if it couldn't claim priority over a handger of goats.

Cardinality, the measure of the size of a set, is very useful because we can use it as a mechanism to compare the size of sets, whether or not we have a numeric value for how big the sets are. If we imagine lining up two sets alongside each other and we can pair off the members of the sets, so that there is a one-to-one correspondence between each pair of members—each member in one set has a unique corresponding member in the other set—then we can say that the sets have the same cardinality, even if we don't know how big the sets are. That last proviso will prove very useful when it comes to dealing with infinity.

To get the idea of using cardinality to compare sizes of sets, imagine, for instance, we had two sets, one being the main compass directions and the other the seasons. We could pair off North with winter, East with spring, South with summer, and West with fall. At this point we've used up all of the directions and all of the seasons, each pairing off with a different element from the other set. So we can say that the two sets have the same cardinality, even if we didn't know how many seasons or directions there were. As it happens, in this case we do know how many. (It's four.) But the important point is that we don't need to know. As long as we can provide a repeating mechanism to make that one-to-one pairing we know that we have sets of the same cardinality.

Remember the oddity that confused Simplicio. Each counting number could be paired off with a square. So we know that the sets of the counting numbers and the squares have the same cardinality—yet we also know that the squares form a subset of the counting numbers. One of Cantor's discoveries would be that it's always possible to find a subset that has the same cardinality as an infinite set.

Before Cantor got to work on sets, they had already been used by the Italian mathematician Giuseppe Peano to define the counting numbers. Back in chapter 2 we derived a number system from real objects—initially only with goats, but eventually applicable to a whole range of objects. What Peano did was to abstract the numbers so they could exist without the objects, based purely on the nature of sets. One of the reasons this approach could not have been taken in the early days of mathematics is that it relies on the fundamental concept of emptiness, zero personified. Specifically, the starting point was the empty set, a set with nothing in it, which was the basis for zero.

The next biggest set was the one that contained just one thing—the already defined empty set. So this provided Peano with the cardinal number 1. He then produced the set containing that set. This contains two sets—the empty set and the set containing the empty set, giving him the cardinal number 2 . . . and so on. The whole set of the "natural numbers" (0 and the positive integers) is built from the ground up this way, using sets that nest like a set of Russian dolls.

Physicist Roger Penrose argues that this ability to define these numbers this way means that they can "seemingly be conjured up by, and certainly accessed by, the mere exercise of our mental imaginations, without any reference to the details of the nature of the physical universe." However, this argument seems to me to incorporate unsupportable sophistry. There is no doubt that we did not "conjure up" the natural numbers by mere exercise of our mental imaginations, and it's hard to imagine that in total absence of physical objects that we ever would.

Even more so, the whole concept of sets as a collection of entities requires the existence of entities in order to be able to conceive of them in the first place—and without countable objects in the real world around us, it's hard to imagine any way that such thinking would ever occur. Say, for a moment, there existed a thinking being that has no physical form and that had no access to the physical world. How would it ever even begin to consider the kind of multiplicity that we experience in the world around us and that produced concepts like the natural numbers and sets, when this being has no experience of anything other than its own oneness?

Set theory, in Peano and Cantor's hands, took arithmetic a step away from counting goats to become the theoretical foundation for basic numbers in mathematics. In one sense this was a powerful abstraction, moving away from counting objects to the essence of numbers themselves. But it was also a move that worried (and continues to worry) some mathematicians, as set theory has a disturbing paradox at its heart. The man who first pointed out the difficulties was British philosopher and mathematician Bertrand Russell.

Just as Peano used sets with other sets as their members to produce the counting numbers, Russell looked at another class of sets that incorporated sets. Specifically he considered sets that did or didn't contain themselves as a member. This sounds painfully recursive, but it becomes clearer with specific examples. Consider, for instance the set "Dogs." This implies an antithesis, the much bigger set of "Everything that is not a dog." Assuming we are happy that we can have an "Everything" that does not just encompass physical objects, then the set "Everything that is not a dog" is a member of itself. Because it's a set, which makes it not a dog. For the same reason, the set "Dogs" is *not* a member of itself.

Here comes the twist that Russell devised. Let's consider the set of "All sets that are not members of themselves." So this is a set that includes the set "Dogs" but doesn't include the set "Everything that is not a dog." Let's call the new set the "NotMembers" set. The question Russell asked is whether or not NotMembers is a member of itself.

By now it's easy to for the brain to have got in an agonizing twist—but that's exactly what Russell had in mind. If NotMembers *is* a member of itself, then it is, by definition, a set that is not a member of itself. Because that's how NotMembers is defined. Similarly, if NotMembers is *not* a member of itself, then it is not a set that is not a member of itself. So it should be a member of itself. It's like trying to deal logically with the statement "This is a lie," which is effectively a phrase that is the equivalent of NotMembers. The wording makes it self-contradictory. In effect, Russell had shown that set theory has an inherent, built-in contradiction—which was not something that endeared it to mathematicians. Yet set theory remained the basis for the nature of numbers and simple arithmetic.

We'll come back to the problems of set theory that Russell uncovered, but first let's take a look at the way that Cantor opened up the implications of infinite sets. If we take Peano's method for constructing the counting numbers to the limit, we end up with an infinite set. It's not the potential infinity represented by the lemniscate, and so Cantor gave it a new label, calling it aleph-null or aleph zero ($\aleph_0$)—the first letter of the Hebrew alphabet with the zero to indicate that this was the basic infinite set. As you might expect from infinity, and as Simplicio discovered in Galileo's book, arithmetic does not work in the way that we are familiar with for aleph-null. It follows from the nature of the set that, for instance:

$\aleph_0 + 1 = \aleph_0$
$\aleph_0 + \aleph_0 = \aleph_0$ and
$\aleph_0 \times \aleph_0 = \aleph_0$

The set theory approach effortlessly does away with the problem that Galileo faced with squares and integers. As we've seen, we can work out if two sets have the same cardinality if the elements of the two sets can be put in one-to-one correspondence. And that's exactly what we do with the integers and the squares, pairing them off, one square per integer. Because we can do this, we know they have the same cardinality—they are both sets of cardinality $\aleph_0$. This makes sense of the odd aspect that an infinite set can be put into one-to-one correspondence with a subset, because as we saw with the compass directions and seasons, it isn't necessary to know how many members are in a set to establish that they have the same cardinality.

The problem comes when we try to understand the mathematical processes in terms of the world that we experience around us. We struggle with the behavior of an infinite set because we expect it to behave like a finite number—specifically like a finite collection of real-world objects. But a set isn't a number, even though it is valuable in understanding them, it is a mathematical construct. And it is only by being clear that we are dealing with an entirely different kind of entity—even if it has relationships that help us understand numbers—that we can cope with an infinite set's strange nature.

So set theory had enabled Cantor to get a hold on the infinity of the counting numbers, $\aleph_0$, and it may seem that this was all that could be said about infinity. But being a mathematician, Cantor was not happy to take anything on trust—and that was the implication of his appending the zero or null part of "aleph-null." This was the basic infinity, but it had not been proved that a set of this cardinality applied to the whole range of numbers beyond the integers. And it was this that Cantor then set out to explore. In doing so he used unusually accessible mathematical proofs. Look at a modern mathematical proof and it is likely to be crammed with page after page of equations. Andrew Wiles's twentieth-century proof of the famous Fermat's Last Theorem stretches to over 100 pages. Yet Cantor's infinity proofs can mostly be appreciated without using any equations.

There is a slight compression in the way we will look at them—the actual mathematical presentations for a solid proof do require more than the conceptual manipulation of simple diagrams, as they have to formalize the process that is undertaken, but the proofs can be understood in terms that are so visual that a Greek geometer would be happy with them. It's unfortunate that some of Cantor's contemporaries didn't feel the same.

The first kind of number Cantor assessed was the rational fractions. He imagined drawing up a table that contained every positive rational fraction by simply adding 1 to the top number of the ratio as you head from left to right in the table, and 1 to the bottom number of the ratio as you head downward from row to row. The outcome is a table like this:

1/1	2/1	3/1	4/1	5/1	6/1	7/1	8/1	9/1	10/1	...
1/2	2/2	3/2	4/2	5/2	6/2	7/2	8/2	9/2	10/2	...
1/3	2/3	3/3	4/3	5/3	6/3	7/3	8/3	9/3	10/3	...
1/4	2/4	3/4	4/4	5/4	6/4	7/4	8/4	9/4	10/4	...
1/5	2/5	3/5	4/5	5/5	6/5	7/5	8/5	9/5	10/5	...
1/6	2/6	3/6	4/6	5/6	6/6	7/6	8/6	9/6	10/6	...
1/7	2/7	3/7	4/7	5/7	6/7	7/7	8/7	9/7	10/7	...
1/8	2/8	3/8	4/8	5/8	6/8	7/8	8/8	9/8	10/8	...
1/9	2/9	3/9	4/9	5/9	6/9	7/9	8/9	9/9	10/9	...
1/10	2/10	3/10	4/10	5/10	6/10	7/10	8/10	9/10	10/10	...
...	...	...	...	...	...	...	...	...	...	...

Clearly, the whole table cannot ever be drawn because it is infinitely large. But we can see how it takes shape. It will contain every possible rational fraction—plus the number 1 represented an infinite set of times, down the diagonal. Now all that

Cantor needed to do to show that the elements of this table had the same cardinality as the counting numbers was to find a simple, repeating path that could be used to step through the table. He needed a set of rules, an algorithm (see page 76) that would enable a straightforward mechanical process of walking through the table, encountering each entry. Something like this:

1. Start at the top left.
2. Take a step right.
3. Move diagonally down and left until you reach the edge of the table.
4. Take a step down.
5. Move diagonally up and right until you reach the edge of the table.
6. Repeat from step 2.

This process will eventually lead you through the whole table, taking in every single rational fraction along the way.

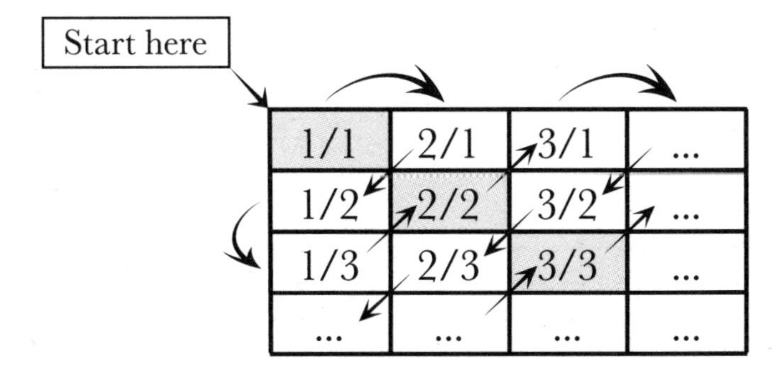

There are other routes that you could take, but the main thing was that Cantor had established a step-by-step route through the table. We've now got one step per entry in the table—and all we need to do is pair those steps off with the integers. There's a one-to-one correspondence step by step, which means that overall we can say that the entries in the table have the same cardinality as the integers. There are $\aleph_0$ rational fractions.

Of itself, this doesn't seem hugely surprising. After all infinity is infinity, and as we know that this equation holds:

$$\aleph_0 \times \aleph_0 = \aleph_0$$

it seems intuitively reasonable that the rational fractions fit the same pattern. However, mathematics is not always intuitively reasonable. And when Cantor took the same approach with another set of numbers he came up with a shockingly different result.

Let's take a look at all the numbers between 0 and 1. (Cantor actually worked with a different range, but this is the simplest one to consider.) What do we mean by "numbers" here? It's not just the integers—there are only two of those, assuming we're using 0 to 1 inclusive. And it's not just the rational fractions, where we would end up with the first column of the table on page 178, a nice $\aleph_0$ subset of the fractions—the ones with 1 on the top and every integer on the bottom. There are also the irrational numbers—numbers like the square root of 2, though here we are looking for all the irrational numbers with values that fall between 0 and 1.

In essence, then, what Cantor was considering was every decimal value (or "real number") between 0 and 1, which should take in *every* possible number in that range. To make this mechanism accessible, we need to scramble the list up, otherwise it will begin with an infinite set of numbers, which have zeroes for as long as you can fit on any piece of paper. Having done that, we end up with something like the table below:

0.**6**4720421758957932691046174...
0.3**1**5729571037472748274373333...
0.61**9**57274945711757299527421...
0.911**3**6728457382356563361221...
0.1622**3**462751225557982367149...
0.23625**0**5585193672152494 7325...
...

Cantor's stroke of genius was to highlight one digit in each number, moving along one decimal place each time. He then added one to each number (if the number was 9, it became 0). This sequence of digits was extracted to make a decimal between 0 and 1. In the case of our table, the number is:

0.720441784983 . . .

Now, this is a very interesting number. It isn't the first number in Cantor's table, because the first decimal place is different. It isn't the second number, because the second decimal place is different. It isn't the third . . . and so on, throughout the table. What we have produced is a number that isn't anywhere in the table.

If we had been able to successfully put every number between 0 and 1 into our list, we could then have paired them off, one by one, with the positive integers, and so would have discovered that this set of decimals had the same cardinality as the infinity of the counting numbers, $\aleph_0$. But in reality, it didn't work out. Cantor had shown (and in his formal version, proved) that there were more numbers between 0 and 1 than there were integers. This set has a higher cardinality. It's a bigger infinity than infinity.

Cantor also took his exploration further, into different dimensions. The cardinality he had just produced, referred to as $\aleph_c$

because it is the cardinality of the continuum of numbers between 0 and 1, was effectively the size of the set of points that exist on a number line, between 0 and 1. However we often describe points on a two-dimensional map, or in three-dimensional space, while mathematicians merrily consider as many hypothetical dimensions as you like. Did the same infinity apply in these cases?

Once more, Cantor's proof can be explained painlessly with hardly any mathematics. If you think of the way a point in a two-dimensional plane is usually identified, it is by a pair of coordinates—the Cartesian coordinates that we have already met (see page 103). These could be x and y on a chart, or latitude and longitude on a map. So all of the points in a 1×1 square can be identified by two real numbers between 0 and 1.

It seems intuitively right that as $\aleph_0 \times \aleph_0 = \aleph_0$, then the same should apply in scaling up the cardinality of numbers between 0 and 1 to cover all the points in a square—but that's not a proof. What Cantor realized was that it was possible to simply interlace the digits of the two numbers that were used to identify the coordinates of a point, producing a single unique decimal that identified that point. And as soon as that was done, it comes down once more to the cardinality of the values that lie between 0 and 1. By interlacing more and more sets of digits, the same applies to as many dimensions as you like. Once more, the infinity of the continuum applies.

Considering the infinity of the numbers between 0 and 1 on the number line produces one of the best of the mind-bending paradoxes that infinity has a tendency to produce. We know that the rational fractions have the same cardinality as the integers, thanks to Cantor's proof. And we're going to use another set of fractions alongside them—the sequence 1/2, 1/4, 1/8,

1/16. . . . It is simple enough to show that these also have the same cardinality. And we already know (see page 39) that the sum of this whole series is just 1. With this established, the fun begins.

Imagine we wanted to protect a positive number line from getting wet. What we are going to do is issue each rational fraction along the line with an umbrella. The umbrella will be a simple T shape. The first umbrella we give out is 1/2 a unit of the number line across the T. The second umbrella is 1/4 of a unit of the number line across and so on. Once every rational fraction has an umbrella, it seems that the whole number line is covered. The umbrella extends half its width in either direction—so, for instance, the first umbrella will cover all numbers for 1/4 of a unit to its left and 1/4 of a unit to its right. Note that this is a rational fraction—and adding it to or subtracting it from the starting point (itself a rational fraction) will reach another rational fraction.

Okay so far? Each umbrella spans from its starting point to a rational fraction on either side of it. Now bearing in mind we've issued an umbrella to each rational fraction, the whole number line is covered, because there's at least a meeting of umbrellas and in most cases they are going to overlap.

We have just covered the whole line from 0 to infinity with our umbrellas. But, remember how wide the umbrellas were. Their widths form the infinite series 1/2 + 1/4 + 1/8 . . . so with no overlaps, the maximum amount of the number line those umbrellas can cover is 1 unit—and with overlaps they will cover even less. A set of items with a width of just 1 covers a line that goes all the way to infinity. Confused? This is the kind of thing infinity does to the brain.

However, despite all Cantor's achievements, one discovery would constantly elude him, perhaps contributing to his eventual descent into madness. Cantor believed that there should be a hierarchy of infinities, of which the infinity of the continuum, $\aleph_c$, was just the first above the basic infinity of the integers. His belief in this structure verged on the religious—in fact he associated the ultimate infinity with the deity. But he could not prove that the infinity of the continuum was aleph one. There may have been another infinity in between. Cantor did not live long enough to discover that his quest was pointless—because it would be mathematically proved that it was not possible to establish whether or not his assertion, called the continuum hypothesis, was true.

The proof was the work of the German-Czech mathematician Kurt Gödel, an implication of his incompleteness theorem. This theorem was any mathematician's nightmare. It demonstrated that whatever your system of mathematics above a very basic level of complexity, it must have some problems that can be formulated in that system that are not possible to solve. The system is the fundamental set of rules underlying the mathematics. Sometimes, even in one world, more than one set of rules can apply, as we've already seen for the behavior of parallel lines on flat and curved surfaces (see page 48). But any attempt to use mathematics to model the real world requires a defined set of rules to be used.

However, at the heart of Gödel's work was the proof that there would always be some problems in any system where the outcome could not be determined. Mathematics was fundamentally imperfect. At a basic level, Gödel's argument was similar to Russell's paradox that we discovered on page 175. It sets up a statement that can't then function properly within the rules of the mathematical system. Gödel managed to show that Cantor's continuum hypothesis was not inconsistent with the axioms of set theory—he couldn't show it was true, but it had the potential to be true. Later, another mathematician, Paul Cohen, would show that set theory was independent of the continuum hypothesis. In other words, set theory would manage just fine if the hypothesis *wasn't* true.

Between them, in essence, they had shown that it was impossible ever to prove whether or not the infinity of the continuum was $\aleph_1$, the next biggest infinity above $\aleph_0$—at least as long as the existing axioms of set theory were unchanged. Gödel himself emphasized this when he said that the continuum hypothesis must be true or false and that "its undecidability from the axioms as known today can only mean that these axioms do not contain a complete description of reality."

As we have already seen, axioms are the fundamental assumptions underlying a branch of mathematics. They provide the "given" starting points that have to be assumed to be true for mathematicians to then build the mathematical structure on them. There are always such axioms in place whenever anything is mathematically proved, though the correctness of the axioms is inevitably a matter of dispute, because in the end they are assumptions, and assumptions can always be challenged.

Set theory, the fundamental basis of all Cantor's infinity work, not to mention the basic concepts of number and arithmetic from the mathematician's viewpoint, relies on a set of axioms known as ZFC named for Zermelo and Fraenkel, two mathematicians who formalized Cantor's work on sets, with the "C" referring to the axiom of Choice (this axiom is singled out for reasons that will become obvious). The eight axioms are reasonably approachable for twentieth-century math:

1. The axiom of existence—there exists at least one set. The cardinal numbers are built from the empty set, the set with nothing in it. But there needs to be something to start with.
2. The axiom of extension—two sets are equal if and only if they have the same members. This is typical of a mathematical axiom in making what appears to be an obvious statement, but one that is required to pin down the mathematics.
3. The axiom of specification—for every set and every condition, there corresponds a set whose members are exactly the same as those members of the original set for which the condition is true. In other words, however you choose some of the members from a set, those members will themselves form a set. So for instance, if we apply the condition "having no positive divisors other than itself and 1" to the set of all natural numbers greater than one, we get another set, the set of prime numbers.
4. The axiom of pairing—for any two sets there exists a set to which they both belong. So effectively you can make a set out of two other sets.
5. The axiom of unions—for every collection of sets there exists a set that contains all the members that belong to at least one of the sets in the collection.
6. The axiom of powers—for each set there exists a collection of sets that contains among its members all the subsets of the given set.
7. The axiom of infinity—there exists a set containing the empty set and the successor of each of its members.

8. The axiom of choice—for every set we can provide a mechanism for choosing one member of any non-empty subset of the set.

Most of these axioms are solid and harmless, but that last, number 8, is the serious problem with ZFC, as soon as we are dealing with infinite sets. The catch is what the "mechanism" can be. Of course, given a set, we could always dip in at random and pick out a member, but "random selection" isn't considered a good enough mathematical method. Although we as human beings could pick a physical object out of a set of objects with no particular reason for choosing it, to select at random using a mathematical mechanism is harder because it is difficult to define how to truly select at random for anything other than a set that has a known number of members.

For finite sets we don't even need to take a random choice. It's always possible to define a mechanism like "take the first member of the set." But with, for instance, the set of all integers, stretching off toward infinity in both negative and positive directions on the number line, how do we define the mechanism for selecting a specific member of this subset of all numbers? It's possible it could be done by the rule "select the median value," but is it obvious what the median—the "middle value"—of an infinite set is?

The good news is that there are now options to "fix" ZFC that would generate a specific answer on the matter of the continuum hypothesis. Most mathematicians agree that it would be best either to use an approach known as "forcing axioms" or another using something known as the "inner-model axiom," which is given the catchy alternative title "V = ultimate L." The only problem is that the forcing axioms approach produces the outcome that the continuum hypothesis is false, while the inner-model axiom makes the continuum hypothesis true.

The two possibilities go to the heart of the nature of mathematics and its relevance to the real world. Set theory is one of the absolute fundamentals of the practical math we use every day, and yet it is based on this arbitrary selection of axiom. Going one route makes the mathematics more exotic and interesting for mathematicians to explore. Going the other makes it closer to the kind of reality we think is the universe we live in. As far as pure mathematics is concerned, there really isn't a problem. There are simply two different mathematical systems at play here—just as there are mathematical systems based on there being thousands of spatial dimensions that have no equivalent in the real world. But for the mathematics we base our science on, we expect a single, definitive approach.

While set theory's problems have a real impact on the application of mathematics to science, there is less of an issue with Cantor's infinity work. We have seen how Newton, Leibniz, and their successors made use of potential infinity in performing calculus to work out equations involving change and values that could be deduced from combining an infinite set of infinitely small slices—and it's impossible to deny the value of infinity, or at least Aristotle's potential infinity, in this context. It is less clear whether or not Cantor's "real" infinity has any relevance beyond the totally abstract world of mathematics.

The simple answer is that as yet we really don't know if infinity has any meaning in the real universe. It is possible that the universe is infinite in expanse. The big bang theory does not preclude this. All we know is that the observable universe is about 90 billion light-years across, a figure combining the distance light could travel in the assumed lifetime of the universe with the expansion the universe has undergone during that time. But it is entirely possible that there is no limit to the expanse of the universe, whether it is considered a single entity or just one bubble in an infinite sea of universes, as many modern big bang models assume that there have been many big bangs in a far bigger multiverse.

In some ways, an infinite universe appears to be more attractive than a finite one, as a finite one begs the question of what happens beyond the boundaries of the universe. However, mathematicians already have one potential answer for that problem—which is that it is perfectly possible for a finite universe to exist that has no boundaries at all.

This might seem counterintuitive, because we have to think beyond the usual three dimensions to properly envisage it, but we can easily envisage a two-dimensional equivalent—the surface of the Moon. (I have chosen the Moon rather than the Earth to avoid the discontinuity provided by oceans.) What we have on the Moon is a finite space—the surface of our satellite—yet it has no boundary. We can continue walking forever in any direction without ever reaching an edge. To picture a similar effect in the universe, just as we fold the apparently flat surface of the Moon (when standing on it) through a third dimension to join up the edges and make it boundless, we have to fold the volume of the universe through a fourth dimension, so that taking a step that would apparently result in leaving one side of the universe would result in entering the opposite side.

Another possibility for a true infinity in the realms of reality is time, which Aristotle imagined would never have an end, and so could be thought of as infinite in expanse. Some cosmological theories have no beginning for the universe either, though the preferred big bang theory gives it a starting point. Some would argue, though, that the universe will end, if eventually the whole universe winds down to the extent that there is no difference from moment to moment. If that is the case, it's possible to argue that time no longer really exists, because there is no way to mark its passing.

Aristotle also believed (he wasn't an atomist) that both time and space could be divided up forever into infinitely small sections. Most physicists now think that this may be untrue—that just as the rest of physical reality is quantized—broken up into discrete particles—so must time and space be, especially if gravity is ever to become part of the quantum framework. The best candidates for a measure of the granularity of space are the Planck length and time, units derived from three fundamental constants, the speed of light, the gravitational constant G, and Planck's constant h.

The idea, when they were first dreamed up, was to define units that weren't artificial and man-made, but were fundamental to the universe. The Planck length is $\sqrt{hG/c^3}$ and comes in at around 1.6×10^{-35} meters, while the Planck time is $\sqrt{hG/c^5}$ making it around 5.4×10^{-44} seconds. There is a third Planck unit that rarely gets a mention, the Planck mass, which at $\sqrt{hc/G}$ is the only one that is comparable with things we have experience of, at around 2.2×10^{-8} kilograms. (I haven't given nonmetric equivalents for these units as such small values aren't ever measured any other way.) It is possible, though certainly not definitively true, that if space and time are quantized, these values could give a measure of their granularity. (The Planck mass is certainly not the smallest possible mass—for comparison, the mass of an electron is around a trillion trillion times smaller. In fact, if anything, it has been argued it could be the largest possible mass for a fundamental particle.)

One interesting possibility of a kind of true infinity emerges from the field of quantum physics. A quantum particle like an atom or an electron has properties that are quite unlike the ones we are familiar with for objects in the "macro" world. If, for instance, you take the property of quantum particles called spin (quantum spin is only analogous to something spinning round, not a literal spin), we can't produce an actual value for it. Instead what is found is that if spin is measured in any particular direction it will always be either "up" or "down" in that direction (hence the quantization).

Before the measurement is made, we can't know what the outcome will be, because the spin of the particle has no true value. Instead the spin is in a state called superposition, where it might, for instance, have a 27 percent probability of being spin up and 73 percent of being spin down. If we repeatedly made the measurement in that particular direction, 27 percent of the time the result would come out up and 73 percent down—but we can only predict the probability (thanks to Schrödinger's equation—see page 214), not the outcome.

In the superposed state, the spin can be thought of as a direction. Think of it as an arrow that points up to represent a 100 percent probability of spin up and down for a 100 percent probability of spin down. In the superposed state, it is as if the arrow points in an intermediate direction. (A 50/50 probability would be the direction at 90 degrees to the up/down direction.) So, in effect, what we have is a representation of a real number, an infinitely long decimal in the form of the direction produced by the superposition of spins. This is why quantum computers, based on such particles, rather than the traditional 0/1 bit of a conventional computer, have the potential to do such impressive work. And if $\aleph_0$ is ever going to intrude into reality, it is perhaps with such quantum bits or qubits that it will become accessible—and even then only indirectly, as we can never read off the value of the direction, but it is real in the sense that it has a direct impact on the result of measurement.

However, one thing that is definitely true is that infinity, which presents an elegant toy for mathematicians to play with, is often a nightmare in science. All too often in physics, as we've already seen with the recoil of an electron, values of infinity turn up. Whether we're dealing with the innards of a black hole or something as everyday as that electron, there is always the possibility for infinity to emerge. Another issue with the electron is that it is thought to be a point particle with no dimensions. But this means that as you get closer and closer to it, its electrical field strength heads off to infinity. And there's nothing closer than the electron is to itself.

The electron's self-energy, the result of it interacting with its own field, is just one of the infinities that plague QED (see page 157), the physics that explains the electromagnetic interaction of light and matter particles that is central to almost all our everyday experience. And yet QED is generally regarded as the most successful theory ever in terms of predicting observed values. How does it get over those infinities? By a mechanism known as renormalization that amounts to replacing ridiculous infinite values with observed ones.

It's not that physicists always have trouble with infinity. The potential infinity that is employed in calculus comes into play all the time. But it is the embarrassing tendency of physical theories to throw up "real" infinities that causes problems. Physicist Max Tegmark argues that by supporting theories that allow for real infinities we are storing up problems for the future of physics.

He points particularly to cosmic inflation, the patch that was applied to fix some issues with the original big bang theory. The idea is that after the big bang, the space of the universe expanded massively, far faster than the speed of light to begin the development of the universe we now observe. And the current version of inflationary theory agrees well with observation. (In a sense, it ought to, because this was how inflation was pragmatically dreamed up, and then changed several times to make it closer to new data. However, for some time now the theory has lined up with many observations that have been made long after the current theory was first produced.)

According to Tegmark, the trouble with inflationary theory is that it allows for a universe that inflates to an infinite volume, which then totally wipes out the ability to make sensible predictions in a wide range of areas, because inflationary theory should, as a result, produce an infinite set of spaces that contain all possible physical situations. When all things are possible, nothing is truly predictable, which undermines the whole point of science. With this viewpoint, inflationary theory is a bit like a killer computer virus that if released unchecked could destroy the whole of scientific theory.

Tegmark has suggested that there should be a limit on inflation's ability to stretch the universe, based on the quantum nature of space-time—just as the finite atoms in a rubber band stop it from being stretched forever, as in principle could be the case if the material was truly continuous. He argues that such a limit would do away with everything from the problems of the infinitely dense black hole singularity to the mathematical issues that get in the way of quantum theories of gravity. We don't need a real infinity is Tegmark's cry.

He concludes: "Our challenge as physicists is to discover this elegant way and the infinity-free equations describing it—the true laws of physics. To start this search in earnest, we need to question infinity. I'm betting that we also need to let go of it." As yet Tegmark stands as something of a maverick, but his thinking could represent a new beginning in science.

Unlike the embarrassing infinities of physics, the mathematics of infinity that Cantor worked on never really had a significant impact on everyday life and science. From our viewpoint of examining the relationship between numbers and reality, what is more interesting out of this whole field is the impact of Gödel's work and the arbitrariness caused by the problems with the axiom of choice. Set theory is at work at the foundation level of mathematics. And yet it has this fascinating flaw. Perhaps more than anything else, these developments show the danger of assuming that reality is directly and literally based on mathematics. Presumably if it were, then reality too would have arbitrariness at its foundation.

However, despite the lack of application of Cantor's infinities, toward the end of his life, physics took on a new direction with mathematics more solidly at its heart than ever. A direction that would change practically everything that was known—and that would transform everyday existence.

13 Twentieth-century Mathematical Mysteries

There is something very strange about the approach taken to teaching physics in schools. All the way up to the end of high school, students learn the same physics as was taught at the end of the nineteenth century. They may have a few tiny peeks past 1900, yet they are unlikely to get any significant teaching of twentieth-century physics (let alone twenty-first) unless they continue to study the subject at university.

It's hard to think that this would happen in other subjects. Imagine, for instance, an English class that considers nothing written since the 1890s—it would be bizarre. And yet literature has changed far less than physics has in this period. Because almost everything that was known before 1900 had to be modified or thrown away as a result of two huge revolutions in the twentieth century: relativity and quantum theory. And each formed part of the new science that had mathematics as an important driver at its heart.

Relativity as a basic concept dates back to Galileo's day, and lovers of symmetry (see next chapter) will point out that relativity is all about invariance, the idea that there is a force or object or property of an object that doesn't change as a result of making a change to its circumstances—the essence of symmetry. Invariance means that something will work exactly the same way, producing exactly the same results, when a specific change is made. In the case of the original, Galilean relativity, that change involved steady motion.

Galileo was famous for trashing a number of the fond beliefs of the ancient world, and relativity does just this. According to legend, Galileo demonstrated his concept during an outing on a lake. He and a number of friends were being rowed at some speed across Lake Piediluco. Galileo asked to borrow a heavy object and his friend Stelluti came up with the key to his house, a large, solid iron object that was irreplaceable. Galileo took the key and hurled it straight up in the air. As the boat was moving forward as fast as six oarsmen could drive it, Stelluti tried to hurl himself off the back of the boat to catch his key as it fell. Galileo's other friends restrained him, and the key fell straight back into Galileo's lap.

What Galileo proposed, and this "experiment" demonstrated, was that when something is in steady motion, like the boat, you can undertake pretty well any physical experiment that doesn't involve interaction with the world beyond the boat and the results will come out just as they would if you weren't moving. If Galileo had been sitting still on the shore, Stelluti would have expected the key to go straight up and back down again. And that, Galileo knew, would also happen on the boat. The "relativity" part was the realization that as far as the key was concerned, the boat wasn't moving. Steady motion can only be defined or detected relative to something else. This is the reason for the "world beyond the boat" proviso. Compared with the shore, for instance, the boat was moving, and it could be easily experimentally detected.

Galileo used no complex math in his workings, yet even Galilean relativity tends to be ignored in school science curricula. The very word "relativity" seems to produce fear in most educators. We teach children the consequences, but we don't really hammer home the crucial nature of relativity. I think most teachers would argue they don't go any further because the subject is too difficult. And certainly when we get to general relativity and quantum theory, the math does reach a level that is well beyond high school students. But there's nothing that is too difficult to grasp in the *concepts* of relativity and quantum theory. In fact, I find when giving talks in schools that children find it easier to get their heads around these than adults, presumably because young people are more used to learning new and weird stuff.

However, Einstein's first move to enhance Galilean relativity, the special theory, is not even particularly mathematically challenging and there really is no excuse for omitting this from the school curriculum. Yes there is math here, and it is different from the basic workings we are used to when dealing with, say, Newton's laws—but it is mathematics that takes no outrageous leaps. And yet this is truly sexy science—the sort of thing that makes students sit up and listen. It's a vehicle for time travel, for example. It makes no sense at all that it is not taught in high school.

In essence what Einstein did was to take Galileo's idea of relativity and to extend it to encompass the nature of light that had emerged from Maxwell's work. As Maxwell had shown, one of the definitive features of electromagnetic waves was their speed. In conventional Galilean relativity, if we move alongside something else that is moving, both of us traveling at the same speed, then as far as we are concerned that "something" isn't moving at all. That's why Galileo could throw Stelluti's key up in the air and have it fall back down in his lap. But if this also held for light, then moving alongside light would also change its speed. And if light didn't travel at the specific speed that defines it, it would cease to exist.

A lesser scientist might have just assumed that Maxwell was wrong, but Einstein was deeply impressed by Maxwell's line of argument and reasoned therefore that, unlike everything else, a beam of light didn't change speed when you moved with respect to it. This seems a relatively small change, yet when it was plugged into the basic mathematics of motion dating back to Galileo and Newton it had a truly surprising effect on the nature of reality.

The simplest example to think of, to see how the impact occurs, is a so-called light clock. This is a device for which each "tick" of the clock is a beam of light flicking back and forth between a pair of horizontal mirrors. The beam travels up and down in a straight line. Let's imagine we had this clock on a transparent spaceship, and had a super telescope that enabled us to see the light clock from Earth. If the ship weren't moving with respect to the Earth, then the people on Earth and the people on the ship would see exactly the same thing when checking out the clock. As the clock was not moving for either of them, the beam of light would travel in a perfect vertical line as it bounced between the two mirrors.

Now let's consider what would happen if the ship was in flight at high constant speed compared to the Earth. Thanks to Galileo we know that as far as the people on the ship are concerned, nothing has changed inside their vessel. The clock is still not moving for them and the light continues its steady up and down straight-line paths. But the people on Earth will see something different. Imagine the beam sets off from the top of the clock. In the time it takes the light to reach the bottom mirror, the ship will have moved sideways. The effect will be that instead of dropping vertically, the light will pass down a longer diagonal path. The same happens on its return journey up to the top of the clock. Now the light is zigzagging in diagonals.

Things would be very different if traditional Galilean relativity had applied. If we imagine watching Galileo's boat on Lake Piediluco from the shore, and he had something like the light clock that fired a stream of ball bearings from top to bottom, then from the boat the "clock" would not be moving and the balls would zip down in a vertical straight line. From the shore we would see both the boat and the balls moving sideways, so we would add together the two motions for the balls to produce a new combined speed. But assuming Einstein was right and light always goes at the same speed however we move with respect to it, we wouldn't see the equivalent from the Earth, looking at the light clock. The light would have farther to travel, but would not be able to change its speed.

This means that something else would have to give way. Something would have to be stretching to allow the light to make its destination on time. When he did the math, Einstein discovered there were three resultant effects. Time slowed down on the ship as seen from the Earth, distances contracted in the direction of movement, and the mass of moving objects went up.

According to special relativity, space and time can no longer be considered entirely separate entities, and it becomes convenient to work with a mash-up of the two, space-time.

As it happens, the mathematics required to cope with this—and it's the reason why special relativity should be accessible to school students—is little more than a touch of ancient Greek geometry and the ability to deal with square roots. Einstein worked out from the geometry of the light clock, and other thought experiments, that the key factor, often given the name gamma (γ) is just $\sqrt{(1 - v^2/c^2)}$ where v is the velocity of the moving object as far as the observer is concerned, in our case the spaceship, and c is the speed of light.

In the case of the light clock, for instance, when a time t has elapsed as far as the spaceship is concerned, for the viewers on Earth the time that has passed by is t/γ, which is $t/\sqrt{(1 - v^2/c^2)}$. That may look like heavier mathematics than you'd like to deal with, but it is pretty straightforward. When v is 0 and the ship isn't moving, then v^2/c^2 is 0, so we just divide t by 1 and end up with t. Both the ship and the Earth see time ticking along at the same speed.

But as v gets closer and closer to the speed of light (c), then the amount that t gets divided by gets smaller and smaller—which means that the time as seen by the people on Earth gets longer and longer. If γ is 1/2, then it's t/1/2—which is the same as 2t. If γ is 1/4, then the elapsed time seen from Earth is t/1/4 or 4t. As far as the people on the ship are concerned, time is passing entirely normal. But from the Earth's point of view, time on the ship is passing slowly, so that when t has elapsed on the ship, 4t has gone by on the Earth.

Perhaps the weirdest thing about all this is that relativity is totally symmetrical. On Earth we tend to take the surface of the Earth as the definition of "not moving," but that is an arbitrary (if usually convenient) decision. After all, the Earth is spinning around its axis, flying around the Sun in its orbit, and moving through space at high speed with the rest of our galaxy. We tend not to think of it that way, but from the viewpoint of Galileo's boat, the boat isn't moving forward, the world is moving backward. Similarly, from the point of view of the people on the spaceship, the ship isn't moving at all. It's Earth that is shooting away from them at high speed. There is no good reason to choose one over the other as the fixed point (or "frame of reference" as scientists tend to call it). So if the passengers on the ship could see a light clock on the Earth, they would see that time on the Earth was running slowly as a result of the Earth's motion away from the ship.

It's not really the point of this discussion to go into all the implications of traveling at near light speed—you'll find a lot more in my book *How to Build a Time Machine*—but it is worth briefly pointing out why this symmetry doesn't hold in the best-known example of this "time dilation" effect. In this thought experiment, a pair of twins are involved in a space mission. One is a mission controller; the other flies off on the spaceship on a long mission at near the speed of light. Let's say both twins are thirty when the mission starts. The astronaut experiences five years passing by, so returns aged thirty-five. But she discovers her twin on Earth has aged ten years and is now forty.

It seems wrong that this should happen if the picture were totally symmetrical as described above. And symmetry would have been in play for the period of the journey when the ship was moving at a steady pace away from the Earth. But then the symmetry would have to be broken. At some point the spaceship had to have a force applied to it to slow it down, then to accelerate it back toward Earth. When it reached home, again a force would have to be applied to match its speed to Earth. It is this change to the ship that doesn't happen to the Earth that effectively resets the clocks so that the ship returns to Earth time, but having only experienced the passage of five years.

There is no doubt that surprising results emerge from the mathematics of special relativity, whether it is time dilation, the way that a massless particle like a photon has momentum or the ultimate equation relating energy and mass, $E = mc^2$. Generally speaking, though, the basic mathematics used is rarely challenging for an advanced high school student. The same can't be said for the general theory of relativity, though. Here even Einstein had to get help with the math, and although the key equations look simple enough, they hide a multidimensional complexity that is challenging enough that they can only be solved for special cases, rather than universally.

The basics behind general relativity, though, which brings acceleration and gravity into the mix (and hence makes this version of relativity more general than special relativity), are reasonably straightforward. Einstein's starting point was what is now called the equivalence principle, based on an idea that he referred to as his "happiest thought."

This thought occurred to Einstein when he was still an amateur scientist, working at the Swiss patent office. (The whole of the special theory was developed here, though the meat of the general theory would come later when he had finally achieved academic recognition.) He later commented: "I was sitting in a chair in the patent office at Bern when all of a sudden a thought occurred to me: 'If a person falls freely he will not feel his own weight.' I was startled. The simple thought made a deep impression on me. It impelled me toward a theory of gravitation."

If you aren't already clued up on general relativity, it isn't immediately obvious what Einstein was getting at in that quote. What he meant is that if, for instance, someone falls off a high building, then he accelerates toward the ground at a standard rate determined by the gravitational pull between the Earth and his body. But at the same time, he feels weightless. He floats around as he falls, just as much as astronauts do in the International Space Station (ISS). That's if you ignore the buffeting caused by the air he is moving through—it's probably better to think of him falling inside a box, with both him and the box in free fall. In fact the only reason that the astronauts on the ISS feel weightless is because they too are falling.

Remarkably, the strength of the Earth's gravitational pull is actually very similar to ground level at the height the space station flies—it's around 0.9 times normal Earth gravity. But the reason the astronauts are effectively weightless is that the space station is plummeting toward the Earth, just as much in free fall as the person who falls off the building. They only difference is that the space station is also moving at high speed sideways, so it keeps missing the Earth as it continues to fall. This is what being in orbit entails: falling and missing.

So those floating astronauts are a real, everyday demonstration of Einstein's observation that people who fall freely don't feel their own weight. As they accelerate downward, the pull of gravity that they should feel is canceled out. By comparison, if we accelerate downward faster than gravity's effect, we feel as if we are in a gravitational field pulling in the opposite direction. Think what happens when you are on a plane and it starts to accelerate along the runway. You feel a gravity-like pressure pushing you back into your seat. If the acceleration is great enough to give the kind of "g-forces" experienced by astronauts or fighter pilots, your body will feel extremely heavy.

So we see what Einstein was getting at. The fact that astronauts and other falling things feel no weight in free fall reflects the way that accelerating and experiencing gravity are exactly equivalent. He had produced, mentally, a similar situation to that envisaged by Galileo. In Galileo's original thoughts on relativity, he imagined being inside the cabin of a steadily moving boat that had no windows. He realized that no experiments that he could do within the cabin would determine whether he was moving or stationary. With special relativity, Einstein added in the invariant nature of the speed of light. Now, bringing in acceleration, he had something more.

Imagine, for instance, that you are inside a spaceship without windows and you feel a constant force pulling you toward the back of the ship, giving you weight. According to Einstein, there is no experiment you can do inside that ship to determine whether your craft is sitting on a planet with just the right force of gravity to pull you toward the back of the ship, or whether the ship is out in space, constantly accelerating under the push of its motors. The gravitational pull and the acceleration are equivalent and indistinguishable.

Pedants (and being pedantic tends to go hand in hand with science) will point out that strictly speaking there is a way to distinguish the two cases. This is because it is possible to move around inside the spaceship. So if you did the experiment twice, once right at the back of the ship, and once right at the front, you would expect to see no difference if the ship were accelerating, because the acceleration would be the same in both places. But you would see a tiny difference under the pull of gravity, as the front of the ship would be farther away from the planet, so you would feel slightly less gravitational force. While strictly true, this misses the point of the equivalence principle, which is about what happens at a chosen point in space, not when wandering around a ship.

While we're on the imaginary ship, we can try out another experiment that is of great significance. Let's imagine we've got the light clock that was used in the special relativity experiment. If you remember, while the ship is moving steadily, as far as the people on the ship are concerned, the clock is not moving and the light continues to travel from top to bottom between the mirrors, undisturbed. But imagine that we now switch on the ship's motors so that the clock isn't moving at a constant speed with respect to the Earth, but is accelerating. Even inside the ship, we are aware of the acceleration. We know that the mirrors are feeling a force that accelerates them while the light is in flight between mirrors. And the light will no longer produce a clear, vertical straight line traveling from mirror to mirror. Instead, as it travels from top mirror to bottom, the people on the ship will see the light following a curved path.

That's interesting in its own right. It makes it clear that Galileo's "no experiment can distinguish" relativity doesn't hold true when there is acceleration. But we didn't need a light experiment to realize this. We can feel the effect on our bodies—or on the accelerometer built into a cell phone. However, the principle of equivalence tells us something much more interesting. We have seen that when the ship is under acceleration it makes the path of light bend. In that case, it must also happen when the ship is in a gravity field. Einstein realized a fundamental aspect of the nature of gravity. It is effectively the ability of matter to warp space-time. The presence of matter with mass makes space-time warp, as a result of which we see the light beam, which is simply trying to head off in a straight line, travel through a curve.

This might sound familiar. Remember the space station orbiting the Earth? It too is traveling in a straight line, at a tangent to the Earth, and should fly off into space. But the mass of the Earth warps space-time and the space station curves into an orbit. This explains why things curve around massive objects as they travel, but it is less obvious why a stationary item, like an apple, starts accelerating toward the ground. The important thing to notice here is that it is not just space that is warped but space-time. And though the apple is stationary in space it isn't stationary in space-time—time is ticking on. So effectively that acceleration is a result of the warping of time.

This gives us a feel for the general theory of relativity at the explanatory level. But no mathematics has yet been involved. It's enough for us to have a picture of what general relativity does, but not enough to make use of it scientifically. One of Einstein's great abilities was to visualize a situation in a thought experiment with limited mathematical content, which he would then populate with the mathematics to give clarity to what the thought experiment seemed to say should happen. In the case of general relativity, that plugging in of the math would take several years, primarily between 1911, when he took a rest from the quantum theory that had been plaguing him, and 1915.

The first problem Einstein faced was to move away from the geometry that Euclid had expounded so clearly a couple of thousand years previously (see chapter 4). As we've seen, one axiom that the Greeks didn't bother to state was that they were undertaking Pythagoras's theorem, and all the other theorems of Euclidian geometry, on a flat surface. That was just taken for granted. However, when you think about it, this was an odd assumption to make. Apart from anything else, in the ancient world there would have been far fewer truly flat surfaces than we have today. While true flatness was okay for Plato's perfection of archetypes, the shadows in the real-world cave were inevitably flawed. It was formally stated that the lines in geometry could have no thickness, unlike the real world, but no one bothered to state the assumption that they were working on a flat surface.

Another reason that it was slightly odd that the Greeks never even noticed they were assuming flatness is that they knew that Earth wasn't. Flat, that is. It's still commonly put about that a flat Earth was the standard model all the way to medieval times, but no one educated in the period thought that the Earth was anything other than a sphere, and that had been demonstrated back in Ancient Greek times. So the geometry of which the Greeks were so proud, a science whose name tells us it was about "measuring the Earth," was actually not appropriate for work on the curved Earth's surface.

We've already seen (see page 48) one way that this is the case. Because parallel lines that run perpendicular to the equator meet at the poles. And it becomes even more obvious when dealing with that most typical of geometric shapes, the triangle. Proposition 32 of Euclid's *Elements* (Book I) tells us: "In any triangle, if one of the sides is produced, then the exterior angle equals the sum of the two interior and opposite angles, and the sum of the three interior angles of the triangle equals two right angles." To put it in a more familiar form, according to Euclid, the angles of a triangle add up to 180 degrees.

In Euclid's elegantly constructed, step-by-step system (remember that he didn't get to this until he'd gone through thirty-one other propositions), there is no escaping this fact; it has to be.

However, it simply isn't true when you move away from a flat surface. On a sphere like the Earth (approximately a sphere, at least), the equivalent of straight lines are called great circles. Clearly no lines can be truly straight in all three dimensions—they bend with the curvature of the Earth. A great circle is any line that, if continued around the Earth would form a circumference of the Earth.

We can do a very simple test with a triangle made with great circles that will blow Euclid out of the water. Imagine we take two points on the great circle that is the equator, some distance apart. From each we draw a line (along another great circle) at 90 degrees to the equator, heading for the North Pole. The two lines meet at some angle at the North Pole. The farther apart the points on the equator are, the bigger the angle at the North Pole. So let's add up the angles of the triangle. We've got two 90 degrees for the corners on the equator, plus whatever the third angle is. These *have* to add up to more than 180 degrees. In fact, if we choose two points on the equator 10,000 kilometers apart (which happens to be the distance from the equator to the North Pole, as that's how the kilometer was originally defined), we get an equilateral triangle. Each side is the same length. And the result is that we have a total of three 90-degree angles—it's a triangle whose angles add up to 270 degrees.

If we make the leap from Euclid to Einstein, it soon became clear that to deal with the curvature of space-time in general relativity, Einstein could not make use of mathematics that assumed the flat realities of Euclidean geometry. His mathematical tools had to deal with curved surfaces. And even more scarily, it had to cope with situations where four dimensions—three of space and one of time—were all curved. Einstein's friend Marcel Grossmann pointed him in the direction of the most advanced geometry of curved space available in his day, developed by the German mathematician Bernhard Riemann.

Einstein's work was by now coming under scrutiny, particularly by the leading German mathematician David Hilbert. This was particularly the case after Hilbert invited Einstein to lecture at Göttingen, Hilbert's university. The mathematician seemed to decide that he ought to finish off Einstein's work, at the time still containing several flaws, before Einstein could. In fact, although Einstein published his first complete paper on general relativity before Hilbert, it was thought for some time that Hilbert had in fact completed the equations first, but published later. More recently it has turned out that Hilbert's original unpublished paper itself had a number of flaws, and was corrected in response to Einstein's paper, so Einstein would probably still have priority.

Whoever practically got there first, there is no doubt that the work was started by Einstein, and there were no recriminations here as there had been between Newton and Leibniz over calculus. Hilbert graciously accepted this was Einstein's baby. The result of all this work was a deceptively simple equation, which can be represented as

$$G_{\mu\nu} + \Lambda g_{\mu\nu} = (8\pi G/c^4)\ T_{\mu\nu}$$

This looks harmless enough, apart from that fourth power on the speed of light, but the elements of the equation with the $\mu\nu$ subscripts are not simple variables, they are tensors, the powerful but hard to manipulate tool that Einstein took from Riemann's work. A tensor compresses a multidimensional relationship between different values, vectors, and other tensors into a single term. Hidden behind the apparent simplicity of that term is typically a matrix of variables. The tensors in the gravitational field equation unpack to produce ten different underlying differential equations, dealing with values that change depending on location in space and time. Underneath that elegant surface, all hell is breaking loose.

When Newton worked out his theory of universal gravitation it implied a simple underlying equation (Newton didn't use it himself) of

$$F = Gm_1m_2/r^2$$

All that is involved is the constant G, the masses of the two bodies and the distance between them (r). That familiar G crops up also in Einstein's equation, but the rest of it had to deal with far more than the influence of mass, important though this remains. One of the key factors was the observation from the special theory of relativity that the mass of an object doesn't have a fixed value, but is modified by movement. Energy contributes to mass—and Einstein would show that it also made a contribution to gravitation. The mass part of the effect primarily produced a warp in time.

Then there was the space warping that started the whole business when we considered the beam of light being twisted in the accelerating spaceship. Unlike time, for space we have to consider each of three dimensions along both directions, effectively bringing six extra factors into play. And finally, Einstein had to incorporate some oddities. One was the factor known as frame dragging, where due to relativistic effects, a moving object generates a small gravitational pull at right angles to its movement. And the final contribution came from the way that gravity generates a form of energy (think, for example, of the potential energy of an object high up in a gravitational field, like a rock on the top of a high mountain). As we've already seen, energy produces gravitation. So there's a small feedback loop where gravity itself is a source of yet more gravity.

Unlike special relativity, general relativity used mathematics that is not, and never will be, familiar to the general public. And cracking these complex equations has proved extremely challenging. Although solutions rapidly began to emerge for special cases, there is no general solution. One of the first of those solutions, like Maxwell's predictions of electromagnetic waves was an early example of a mathematical prediction being used to bring an unexpected physical entity into being—in this case, the black hole.

As a concept, the black hole dates back to the eighteenth century, when the English astronomer John Michell realized that if a body were massive enough, its escape velocity—the speed you have to throw something for it to escape the pull of the body—would be so high that it would exceed the speed of light. Even light would be trapped by such a dark star. But Michell's work was long forgotten when the German physicist Karl Schwarzschild revisited the concept as a response to Einstein's theory.

This was in 1916, when Schwarzschild was fighting in the First World War. Just months after the publication of his theory, Einstein had only managed an approximate partial solution, making a prediction that could be tested about a variation in the orbit of Mercury. But somehow, Schwarzschild found time in the trenches to produce an exact solution for the special case of the black hole, a star that had collapsed effectively to nothing, unable to resist its own gravitational pull as the fuel that had been puffing it up ran out.

Schwarzschild didn't call his solution a black hole—the name would not crop up until the 1960s. It is usually ascribed to John

Wheeler, who was certainly one of the first to use it. However, he doesn't seem to have been the first to think of it. The term was bandied about by someone other than Wheeler, at an American Association for the Advancement of Science meeting in January 1964, and as a result appeared in print in a *Science News Letter* article about the event written by Ann Ewing. It came out of the meeting, but no one knows for sure who first used it.

Wherever the name came from, we know where black holes came from: from mathematics. The way we test the effectiveness of a scientific model tends to be in its ability to predict real-world outcomes that weren't used to create the model in the first place. This is why Einstein was so enthusiastic about producing a partial solution to predict how Mercury's orbit should behave, and soon after, the predictions of his theory about gravity bending a beam of light would be tested by observing stars whose light passed close to the Sun, visible during a solar eclipse. But the prediction of the black hole was one of the first examples of a model predicting the existence of a whole new entity that no one had seen or was even looking for.

Even now, black holes are more the product of mathematics than of science. Astronomers have observed many objects in deep space that behave from indirect evidence as if they were black holes. The evidence is strong, but remains indirect. We have never observed a black hole. Instead we have a prediction from the mathematics that we would expect to produce the kind of effects that we see, for instance, at the center of our galaxy where it is suspected there is a supermassive black hole. In terms of the overall thesis of this book, this is mathematics absolutely on the borderline of reality. It seems to have a bearing on the real universe, but that has yet to be proved and some, like Max Tegmark (see page 192) suggest that the link between the math and reality will never be good enough—that the predictions will prove wrong if we ever get to analyze what appears to be a black hole up close.

Important though relativity was to prove, there was another equally transforming branch of physics emerging at the same time, in which Einstein also had a significant hand—quantum physics, the physics of the very small. At the start of the twentieth century, atoms were not considered definite existing entities, but rather they were seen as a useful conceptual tool to help predict how matter would behave. But with the early years of the century it was demonstrated not just that atoms did, indeed, exist—Einstein was primarily responsible for that demonstration—but that they behaved in ways that seemed entirely unnatural, a true paradox since most of nature was based on such quantum particles.

This isn't the place to detail the complexities of quantum physics—the study of the universe on the scale of tiny particles like atoms, electrons, and photons of light—but there are two key observations from the way it was developed that are important in understanding the impact of mathematics on our view of the universe. The early workers in the quantum field like Einstein and the Danish physicist Niels Bohr quickly realized that they would have to throw away many assumptions about the way that matter and light behaved when looking at thc world at the quantum level. Light, for instance, had clearly been established to be a wave, and Maxwell's work seemed to confirm this. But the new quantum sages would show that light's wave behavior was nothing more than a model. Light had wave-like properties, certainly. But it could equally be described as a collection of particles or as a disturbance in a field.

Bohr tried to explain the way that electrons in an atom interacted with light by placing those electrons in orbits, like planets around the Sun, a model that we still see in almost all graphic design representations of the atom, even though this model was out of date before the 1920s. Bohr quickly realized that such orbits were impractical and instead put the electrons, in effect, on rails, able to jump between sets of rails, but not to exist in the space in between. "Real" objects that we are familiar with at the level of things we can touch and see don't behave like this, but these quantum particles seemed to do so.

Bohr's work on the atom was limited in its application, and rising young scientists, notably Werner Heisenberg and Erwin Schrödinger, took on the task of extending the mathematical description of quantum interactions far beyond Bohr's simple atomic model. Heisenberg in particular took a purely mathematical approach, developing a description of the behavior of quantum particles called matrix mechanics that made no attempt to take into account a model that could be visualized of what was happening. Instead he manipulated matrices—arrays of numbers—turned the handle, and this black box model, like that eventually used by Maxwell on electromagnetism, turned out useful predictions.

Schrödinger was less happy with such an abstract approach and instead took a parallel look at modeling quantum behavior by using the more familiar form of waves. His equation, still a key component of understanding quantum physics, is another that looks a lot simpler in its commonly used form than it truly is

$$i\hbar\frac{\partial}{\partial t}\Psi = \hat{H}\Psi$$

There are a number of reasons why this is mathematically scary. One is the presence of i, the square root of –1 (see page

97). And rather like the tensors in Einstein's equation, the symbols Ψ (Greek letter psi) and H with a hat on here are not what they seem. Psi is the wave function, a complex formula that describes the nature of a particular system, and the H-hat is a "Hamiltonian operator" a shorthand way of representing the energy in a system and applying it to the wave function.

Initially it was thought that the wave function in Schrödinger's equation, or rather the square of its value (thankfully losing that worrying i) represented the location of particles in the system, but if that were the case, it seemed to predict that all quantum particles spread out over time, getting bigger and bigger, for which there was no experimental evidence. (Thankfully—it would make for a very strange universe.) It was realized that, instead, the equation predicted the *probability* of finding a particle in a particular location or a system in a particular state.

Quantum theory went on far beyond just this equation and in its use of convoluted mathematics. There was, for instance, the Dirac equation, bringing in relativity to the behavior of quantum particles (and predicting as a side effect the existence of antimatter), the extension of quantum theory to quantum electrodynamics, predicting the interaction of light and matter, and more. But we only have to get to Heisenberg and Schrödinger to see the two big steps forward in the ability of scientists to "believe" what was coming out of the mathematics, rather than observation.

First, Heisenberg produced a totally mathematically oriented description in his matrix mechanics. Then Schrödinger's equation brought probability into the mix. This was the point where Einstein parted company with the theory he had done so much to bring into being. Einstein hated the idea that, at the level of quantum particles, when nothing is being measured, all that exists is probability. A quantum particle doesn't have a location unless it has just been measured, for instance. It is often stated that quantum particles can be in two places at once, but the reality appears to be far stranger. All that exists is the web of probability.

In essence, quantum physics appears to make reality fundamentally mathematical. According to the theory, there is no real world, no observable equivalent to this dependence on probability, when measurements are not being made. It might seem at first sight that this is just a variant on what happens when we toss a coin. In a fair coin toss we usually say that there is a 50/50 chance of the outcome being heads or tails. But in reality, once the coin has been tossed, there is a 100 percent chance of one outcome and a 0 percent chance of the other. We just don't happen to know which that value is until the coin is revealed. But if we had a quantum coin, there would literally only be the 50/50 probability with no underlying certainty until a measurement is made and the result is produced.

While this relationship between a mathematical concept and real quantum entities seems unlikely, it has been confirmed by a great many experiments over the years. However, it turns out to be an exaggeration to say that it is truly a case of the real world being based on mathematics. In a sense, it isn't as surprising as it first seems. We are still dealing with models of reality, rather than an absolute description. And probability is rather different from the abstract mathematics that so often drives modern physics, in that it was developed all along as an applied mathematical tool—it derives from an observation of real operations like coin tosses, it's just that in the quantum form, it takes on an independence that was never there in the original formulation. It doesn't mean that the quantum particle *is* a probability, merely that all we can usefully say about it is in the form of probabilities.

Even in the time of Galileo and Newton, symmetry kept cropping up in physics, whether it was in the pleasing reflective symmetry of Newton's third law of motion, or the way that different views in relativity have a symmetrical impact. But as the twentieth century rolled on, symmetry began to play a new and dramatic role in the shaping of scientific theories. The math was truly taking over.

14 Symmetry Games

Science and mathematics have yet to produce many great female role models—not because women are worse at the subjects, but simply because they were for so long excluded by a bizarre culture that was sure that anything so mentally challenging would cause their little brains to overheat. Ask anyone to name a great female figure in math or science before the second half of the twentieth century and they might mention Marie Curie, Ada Lovelace, or maybe, at a push, Caroline Herschel, but they are unlikely to pick out Emmy Noether. Yet this German mathematician was single-handedly responsible for the way that mathematics began to set the agenda in physics. Specifically, Noether identified the importance of symmetry, discovering that it wasn't just easy for nature to produce symmetrical structures, but that the mathematics of symmetry appears to be behind many of the physical laws.

More often than not, math now leads the way and an attempt

to obtain a match to reality follows. Noether's discovery that symmetry produced the laws of conservation was directly responsible for completing this shift, started by Maxwell. Our standard model of the fundamental particles that make everything up does not originate from fancy experiments like the Large Hadron Collider, but from the mathematics of Noether and her successors. And in many ways this has proved remarkably successful. Yet sometimes it seems as if the whole process has gone too far.

It is probably unfair to say that the only reason Emmy Noether is an unfamiliar name is because she was a woman. If you were to ask the general public to name *any* twentieth-century mathematicians, it's likely that they couldn't come up with a single name who wasn't actually a scientist rather than a mathematician—in fact given the amount of publicity Noether has received recently, she may well be one of the few who has a possibility of being named. Yet because of her work's importance to the way physics has developed, it's arguable she should be as well known as Schrödinger and Heisenberg.

Born in Bavaria in 1882, Noether was the daughter of a respected mathematician, Max Noether. Her parents named her Amalie, but from a young age Noether called herself Emmy, which was soon accepted by everyone. Unlike many great mathematicians and scientists, Noether showed no great interest in the subjects at school when growing up—she first qualified to teach languages and it wasn't until 1903 that she registered as a student at the University of Erlangen, gaining a PhD in mathematics in 1907, only the second German woman to do so.

In 1909 she moved to the University of Göttingen at the suggestion of Einstein's challenger, David Hilbert, and it was here in 1915 that Hilbert put Noether forward for habilitation, the unique German requirement of producing a kind of second post-doctoral thesis to become a professor. This was not an option legally available for women, so Hilbert and his supporters petitioned the government to make a special exception in Noether's case. The original petition was rejected and it wasn't until 1919 that Noether was allowed to habilitate, though even then she did not become a professor. Like many mathematicians, she did what is generally considered her most important work while still young. Noether's theorem, which we will return to shortly, was devised early in her time at Göttingen. She did make a number of other very significant mathematical discoveries, but none that would have the same impact outside pure mathematics.

Things went relatively smoothly at the university until 1933, when Noether was removed from her position by the newly ascendant Nazi party. This has often been linked to her Jewish ancestry, and that may have been a contributing factor, but like a number of her colleagues who weren't Jewish but were also expelled, she had communist sympathies and had taken on a guest professorship in Moscow in 1928—it is more likely that it was this connection that proved her immediate downfall. Noether attempted to return to Moscow, but this proved difficult in those turbulent times, so she moved to the United States where she died in 1935, age just fifty-three.

To understand Noether's theorem and why it has become so important to theoretical physicists, we need to know what symmetry is, why it produces these remarkable results and how far this approach stretches the relationship between mathematics and reality. In ordinary English usage, when we talk of symmetry we mean reflection symmetry—the kind of symmetry that we see in a mirror image. Many living things have this kind of symmetry, which can be recognized when the organism has a "left" and "right" side that are pretty much mirror reflections of each other. In fact, this symmetrical nature is central to our idea of human beauty.

Beauty may be in the eye of the beholder, but there are certain features that we look for in, say, a beautiful face, and symmetry appears to be one of the most important. Repeated testing has shown that humans—and, strangely, chickens—consider humans with the most symmetrical faces to be more attractive than those whose faces are less symmetrical. It is thought that this may be because a pronounced lack of symmetry is often produced by illness, and if beauty is primarily an indicator of a potential mate, then lack of symmetry may mean reduced reproductive capability. Searching for symmetry may make sense as an evolutionary trait.

Mathematicians recognize a whole swath of symmetries as well as the simple mirror variety. Generally speaking, a symmetry is present if we can make a change in an object and after that change, the result is indistinguishable from the original. So with left/right mirror symmetry, we can swap the left- and right-hand sides of the image and make no change to the outcome. It can be easier to think of a simple shape to illustrate symmetry: if you imagine swapping the left and right halves of a rectangle or a square, we get that indistinguishable mirror outcome.

Another simple possibility is symmetry of rotation. Take the square and, unlike the rectangle if you rotate it through 90 degrees, once more it is indistinguishable from the original. But rotate it through, say, 45 degrees and this is not the case. We can tell that something has happened. The appearance of the object changes. Compare this with a circle, which has the highest level of rotational symmetry possible. You can spin a circle around by however many degrees you like, or even by fractions of degrees, and the outcome is unchanged.

To go beyond these simple symmetries, we have to slightly extend the way that symmetry is applied. For example, we could consider symmetry in time. If the view we see is unchanged after a period of time has elapsed, then it can be considered symmetrical in time during that period. A totally static object has the maximum symmetry through time, as it always looks the same, while an object undergoing a cyclical change—the second hand of a watch, for instance—has a limited symmetry through time, rather like the rotational symmetry of a square.

It is in thinking of symmetry through time that we see the most obvious example of the way that relativity has an influence on symmetry. It's one thing to say that a totally static object has maximum symmetry, but the concept "static" is a relativistic one. It is not an absolute. If you imagine the symmetry through time of a spaceship, for instance, it has that maximum symmetry due to not moving from the viewpoint of an observer on the ship. To them there is no change in its appearance (limiting our comparison to the outer hull and excluding any wear and tear). But to an external observer who sees the ship shoots past, the symmetry through time is clearly broken.

Another kind of symmetry that needs a mathematician's brain to be totally comfortable with is translational symmetry. Here we take time out of the equation, but just consider two snapshots and ask if the outcome is the same. For example, if I take my square and move it one square's width in space to the right, then after doing this, it is obvious in the real world that the outcome isn't truly symmetrical. I can see that the square has moved. But what mathematicians do instead is to imagine that the square forms part of an infinite surface on which it occurs in a repeating pattern. They then ask, could you tell the difference if this sideways move took place? and the answer is "No"—so there is a translational symmetry in the pure mathematical world. However, if the square moved sideways by a distance involving a fraction of its width, rather than the whole width, the result would not look unchanged and the symmetry would not exist.

There is no doubt that nature often approximates to symmetry. Plenty of physical structures have a near-symmetrical form, not because symmetry has some magical power, but simply because it is an efficient way to grow, or because the forces forming the structure are the same in all directions. Most animals have some kind of symmetry, whether it's the bilateral, side-to-side mirror symmetry we have already met, or the more exotic rotational symmetry of a starfish. Raindrops and eggs and blades of grass and stars all exhibit symmetry, or at least its Platonic shadow form, as in reality these things are not perfect in their symmetry—arguably the only "real" thing that can achieve this is a black hole's event horizon . . . assuming black holes are real (see page 212).

What's more, there is a lot of symmetry in physics. Not just the reflecting symmetry of Newton's third law with its "for every action, there is an *equal and opposite* reaction," but also in the way that you can apply shifts in time and space, or rotation, and still find that exactly the same rules apply. The universe would certainly be an interesting place if, for instance, the relationship between force and acceleration changed when we turned around and faced in a different direction, but such inconsistency would make science hard to deal with in a practical manner.

Similarly, we have to assume that physical laws remain

unchanged as a result of shifts in position and time, because we couldn't do physics unless we made this assumption. Thankfully, experience to date makes it seem likely that this is the case, though it can never be proved. If the force of gravity were constantly changing, it would be impossible to make any sense of the physics involving gravity. Almost all our scientific models start with an axiom that the laws and constants of physics are symmetrical in this fashion. This is a purely pragmatic decision. For instance, pretty well all cosmology and astronomy would be messed up if the speed of light were constantly changing, because we make the assumption of its constant speed to make deductions about how far we are looking into the past when we see a distant object.

Occasionally iconoclasts will make tweaks to this axiom. So, for instance, physicists Andreas Albrecht and João Magueijo have suggested that the speed of light has shifted significantly in the lifetime of the universe, as a result of which it would not be necessary to introduce the concept of inflation, which requires the universe to have gone through a sudden and unexplained phase of massive expansion to allow areas of the universe to have come into equilibrium that are now too far apart to have been in contact during the assumed lifetime of the universe. But mostly the axiom remains unchallenged, because without that assumption it becomes pretty well impossible to do physics.

The approach taken is a bit like the old scientists' joke about someone searching for the keys that he dropped on his way home. The person knows that the keys were lost before he reached the street he lives in. Yet he spends all his time looking in his own street. "Why?" asks a friend. "You'll never find the keys here." The answer is simple. Because the street he lives in is the only one on his route home with streetlamps. As it's dark, there's no point looking anywhere else. Similarly, while the assumption that laws and constants don't change with shifts in time and space may be entirely wrong, there is no point trying to do science outside the "light" of this axiom.

Having said that, we do have evidence that shows at least some physical constants and laws are symmetrical through time, giving more weight to the assumption. A good example would be the size of the electrical charge. We can check this is true going back a good 2 billion years because of some remarkable natural nuclear reactors that were discovered in 1972 at Oklo in Gabon, West Africa. Back when these natural reactors were formed, there was significantly more of the uranium-238 needed for fission reactors, enough that where the isotope was particularly concentrated at Oklo, a nuclear chain reaction began, pouring out heat and radiation.

As it happens, one of the products of this reaction, samarium, could not have formed had the mass of its nucleus been slightly different. Yet we know that the mass of the nucleus is somewhat dependent on the charges of the protons confined in the nucleus, because of the equivalence of mass and energy. And from the samarium found in these ancient reactors, it is possible to deduce that the electric charge we measure on a proton could not have been more than 1 in 10 million different from what it is today, or the samarium would not exist. This kind of measurement makes it possible to establish that at least some familiar laws and constants have remained pretty much unchanged over billions of years, though it is worth noting that any variation would most likely have happened in the very early days of the universe, over 13 billion years ago (and even that number is dependent on these assumptions), rather than a mere 2 billion years back.

What Emmy Noether proved with her breakthrough theorem was that there are unbreakable links between symmetry and conservation. If it turns out that the laws of physics are unchanged with a translation in time, for instance, then it follows that energy has to be conserved in a closed system. And, for that matter, if we discover that energy is being conserved, then the laws of physics have to be symmetrical in time. The same applies to other symmetries. If there is no change in the way a system acts when it is rotated, then angular momentum must be conserved. If translation through space makes no difference, it means that linear momentum is conserved.

Not only did symmetry prove to be a way of understanding the reason for the existence of conservation laws that had been until that point something of an assumption, it enabled new directions to be taken, particularly once quantum field theories like QED transformed the physics of the very small. And symmetry would be behind the discovery of new levels of complexity in the nature of the particles that seem to form the basis of everything around us.

What has all this to do with the roiling, random, confusing mess that appears to be going on at the level of quantum particles? Physicists need simplification to understand and model reality. And symmetry offers a significant path to making that simplification, even if it is at a cost of impenetrable math. This move started off with the discovery of the neutron in the nucleus. When British physicist James Chadwick proved the existence of a particle that Ernest Rutherford had predicted some while earlier, a particle that exists in most atomic nuclei that has no electrical charge, it extended the reach of whatever unknown force (what we now call the strong force) was holding protons together.

Protons are positively charged, and when collected together, as is always the case with like electrical charges, they repel each other. Because they are so close together in the nucleus, the repulsive force between them has to be huge, which implies that there is an even stronger force keeping them together. Now, with Chadwick's discovery, it seemed that this force also attracted the uncharged neutrons. It wasn't obvious that this would be the case before neutrons were discovered. If there had only been protons in the nucleus, there could have been an opposite equivalent of the electrical charge, which meant that like charges were attracted to each other. But this force also worked on neutrons, making it something completely different. The man to give symmetry a central role in explaining the behavior of the nucleus was Werner Heisenberg.

As we saw in the previous chapter, Heisenberg did not have a problem using mathematical models for quantum interactions that had no equivalent in the macro world, with no real-world analogy to understand how or why the model worked. The numbers in his matrices matched what was observed and as far as he was concerned, nothing more was required. Now he made the audacious step of using a symmetry that didn't really exist to work on the nucleus. Instead, what he made use of was a "sort of symmetry," a near miss that seemed close enough to imply that there was something important behind it.

Heisenberg noted that protons and neutrons had very similar mass, differing by only around 0.14 percent. What's more, atoms tended to have roughly similar numbers of protons and neutrons in their nucleus. Admittedly there were plenty where the numbers weren't the same. Hydrogen, for instance, the simplest atom, has just 1 proton and no neutrons at all, while the main isotope of the simple element lithium, lithium-7, has 3 protons and 4 neutrons. However, generally speaking, the most stable nuclei are those with similar numbers of protons and neutrons.

It seemed reasonable to Heisenberg, inspired by a sense of symmetry (though not by actual data), that the force holding the nuclear particles together had the same strength for protons and neutrons. Heisenberg imagined turning neutrons into protons and vice versa—in an atom with the same number of each in the nucleus, there would be no change in the situation. There seemed to be a kind of symmetry here. Heisenberg called the symmetry isospin for no obvious reason apart from the opportunity to cause confusion, as the concept has nothing to do with spin. By assuming that protons and neutrons had equal and opposite isospin charges (+1/2 for the proton and –1/2 for the neutron), Heisenberg was able to make predictions about nuclei that were matched by reality.

As various other particles were discovered, the American Murray Gell-Mann, extended the concept of isospin with a second dimension of symmetry that he called strangeness. The result was that the particles fell in neat arrays of differing strangeness and isospin—when plotted out on a chart, they formed groups of eight in an appealing symmetrical fashion. Such a symmetry suggested that there was some underlying cause for their structure, some connecting feature that makes the symmetry emerge.

Yet though there was a good symmetry among the particles in terms of the assumed isospin and strangeness charges, there was a problem because this was all based on increasingly flaky reasoning. The symmetry was imperfect to begin with. The neutron and proton *don't* have the same mass, for instance. Things were even worse when the other particles were brought in. Particles known as "cascade particles" that formed part of the same group as neutrons and protons had masses that were about

40 percent greater than their fellows. If there really was symmetry here, it had somehow been broken. Yet the concept—the mathematics—seemed too appealing to let it go.

Gell-Mann looked for an underlying structure the particles could share that could produce this kind of symmetry, and the most obvious cause was that particles that were assumed to be fundamental like the proton and the neutron had subcomponents, which were responsible for the way that the charges were distributed. Just as the atom was made up of smaller particles, so it seemed that protons and neutrons and the zoo of new particles being discovered in cosmic rays and particle accelerators were also made up of smaller particles, working together to produce the varying electrical, isospin, and strangeness charges that were observed or assumed.

We now call those subcomponents quarks, the name that Gell-Mann chose for them. It is often assumed that he took the name from the James Joyce "novel" *Finnegans Wake,* which features the line "Three quarks for Muster Mark!" However, Gell-Mann himself said that the name had come to him first as a sound—*quork*—and it was only afterward that he linked the sound to the phrase and adopted the spelling that has resulted in quarks being pretty much universally mispronounced. However the word is spoken, it's a more interesting name than "aces," which was given to a similar concept developed independently by George Zweig at CERN. In the end it was "quark" that stuck.

Just as Planck had first suggested the photon as a theoretical concept to make the mathematics work, Gell-Mann did not really see quarks as particles that truly existed, simply as a useful way to simplify the mathematical structures, but now they are accepted as true particles that are (probably) fundamental. Gell-Mann originally envisaged quarks as coming in three "flavors": up, down, and strange (hence the effectiveness of the Joyce quote). By combining these in pairs and triplets, the various observed particles could be produced. What's more, by assigning small differences in the mass of the up and down quark, which made up protons and neutrons, and a significantly larger mass to the strange quark, the reason for the symmetry between the particles being broken is explained. There is a true symmetry of fundamental *particles* underlying the observed particles, but not a symmetry of *masses.*

As it happened, this simple set of flavors was not enough. The more observations were made, the messier the situation seemed to be. The outcome, after around twenty years, was quantum chromodynamics (QCD), named by Gell-Mann by analogy with the successful quantum electrodynamics (QED) theory that described the interaction of light and matter. QCD assumed that the force between quarks, carried by particles called gluons, is not a simple one, but rather one that comes in a range of "colors" (nothing to do with actual colors, just as the flavors aren't really flavors)—red, green, and blue. In effect, each quark came in three different colors, and the antiparticles of the quarks had complimentary colors, known as anti-red, anti-green, and anti-blue.

The clever part of this system was that quarks always combined in such a way to make white (or at least in the combination that would be white if these were lights). So particles like protons and neutrons made up of three quarks always had to have one each of the red, green, and blue flavors, while particles made up of two quarks, like mesons, had to have one each of a color and its anticolor. To make this work, there had to be color-specific gluons as well—eight of them in total. The neat thing about this approach from the mathematical viewpoint is that it rebuilt the symmetry around this "color charge." While the symmetry with quarks was inevitably incomplete because of their different masses, the symmetry among the massless gluons was perfect.

The more theoretical physicists and applied mathematicians delved into symmetry, the more it seemed to provide the foundation of the universe. Symmetry seemed to be a natural tendency, which meant that the mathematics of symmetry were applied more and more to make deductions about the nature of reality. However, there was a problem. We have already seen a lack of true symmetry in the masses of particles. And when we look at the different forces of nature, they are hugely different. It would seem that if there were a true underlying symmetry it had to have been broken in a big way. The emerging models that were being deployed to represent the early start of the universe implied that it began in total symmetry, which was now long gone. But how had that happened?

This produced a problem that is a bit like the axiom of choice in set theory (see page 187). It is easy enough to see how to make the choice of an element with human intervention, but how do you make it without someone to do the choosing? Similarly, if symmetry had once existed and had then been broken, what had caused the symmetry to break? The need was to find some mechanism for "spontaneous symmetry breaking." A picture that is often used to illustrate this is a pencil that is balanced on its tip. The pencil will inevitably fall over, breaking the symmetry of its upright position by now pointing in one direction or another. But we can't predict which way it will fall from its initial condition.

Unfortunately, this is a flawed picture. If the pencil truly had perfect symmetry, it would never have fallen, because there would be no force to start the movement. Instead, like the spinner in the movie *Inception,* the pencil would remain in an apparently unnatural upright position. It's only a small lack of symmetry, either in the way the pencil was balanced, the shape of the pencil tip, or any subsequent forces applied to it like an air current, that will result in it falling. We expect it to fall because we live in an asymmetrical world.

As we will see, symmetry is being used widely as a mechanism to drive our understanding of the universe, making it a prime mechanism to generate physical theories, but we do have to be careful. Nobel Prize winner Leon Lederman points out when describing how the assumption of symmetries being real can result in the construction of misleading scientific models:

> Symmetry can be a powerful tool, even when it is only an approximation to reality. But our human species has often made mistakes, assuming some things have or are perfect symmetries when the symmetries are actually only illusory or accidental consequences of something else.

Nonetheless, despite this flaw, we certainly get spontaneous near-symmetry breaking and it is thought that this has resulted in the variety of forces, for example, that we see today. This tends to happen when a system goes from a relatively high-energy state to a lower one. The high-energy state is more likely to be randomized, and hence will be more symmetric. The pencil standing on its end has more potential energy than when it is lying on the desk. Similarly, a traditional magnet loses its magnetism when heated above a certain point because it no longer has aligned magnetic domains within it—the kinetic energy of heat has randomized them. By driving understanding from a mathematical model it was possible to produce a picture that unified the weak and the electromagnetic forces, a situation that appears to have been a reality in the early days of the universe, but as the universe cooled, spontaneous symmetry breaking resulted in the two forces splitting.

However, elegant though the mathematics was, it did not actually fit observed reality. The symmetry approach required the boson particles that communicated forces, like photons for the electromagnetic force and gluons for the strong force, to have no mass. And this was the case in those two examples. But the third force, the weak force that crops up in nuclear fission and is responsible for changing one type of particle into another, was carried by three different particles, all of which did have mass. And that seemed to throw the whole symmetry idea out of the window.

Because this loss of the power of symmetry seemed unacceptable to those whose physical outlook was entirely driven by mathematics, there had to be a fix that would enable symmetry to continue its rule. And so a truly outrageous suggestion was made. What if those weak force carrier particles were indeed massless, but there was an extra force in nature, produced by a field that ran throughout the universe, just as the electromagnetic field (and others) appeared to do? This would be a very peculiar field whose only real role would be to provide a kind of drag on particles, giving the bosons that carried the weak force the illusion of having mass. The field was given the name of one of the (several) developers of the theory, becoming known as the Higgs field.

With the patch of the Higgs field added, it was possible to bring a hidden symmetry back into the process. The only problem was that there was no evidence for this Higgs field. It was an arbitrary fix to theory without any experimental evidence to back it up. Hence the importance of the search for the Higgs boson, the carrier particle of the Higgs force. And, as the news media delighted in getting confused over in 2013, the Large Hadron Collider at CERN produced results that were compatible with the existence of the Higgs boson—though it should be emphasized that this was all very indirect confirmation, especially as the theory could not predict what mass the Higgs particle would have.

We are in the position now where we have a mathematically derived model that has achieved a wide range of real successes. The standard model of particle physics, based on the symmetry approach described, has been very successful at making predictions that matched reality, despite a few remaining issues. However, the model has a lot that is plugged in from observation rather than predicted by the underlying structure. As yet the model can neither explain the nature of the dark matter that is thought to be more prevalent in the universe than ordinary matter, nor does it explain why particular symmetries and masses apply—they just do. And there is no obvious relationship between the very different matter and force carrier particles.

There have been attempts to get around some of these problems, mostly heavily driven once more by mathematics. So the joy of symmetry has led some to suggest that there is a whole new level of symmetry, so-called supersymmetry, which links those two broad families of particles. The only problem is that this is a "simplification" that would make the standard model far more complex, because every particle would need to have a supersymmetric partner that is of the opposite kind. Force-carrying bosons, such as the photon and gluon, would need matter-type equivalents called photinos and gluinos. At the same time, matter-like fermions, such the electron and quark, would need force-carrier equivalents, known as the selectron and the squark.

There is no evidence for any of these supersymmetric particles existing at the time of writing. If this theory were correct (and there is no particularly good reason why it should be), then in a perfect symmetrical world, particles and their superpartners would have the same mass—so it should be easy to detect, for instance, selectrons. As they haven't been detected to date, symmetry needs to be well and truly broken here, pushing the masses of the superpartners up to or beyond that of the Higgs. This means that one of the targets of future, higher energy runs of the Large Hadron Collider should be able to increase, or more likely decrease, the chances of supersymmetry being useful.

Supersymmetry is only a starting point, though. This rabbit-out-of-a-hat production of theory from pure mathematics has gone even further with string theory, which adds whole new layers to "simple" supersymmetry. String theory is one of those ideas that sounds beautifully simple when described at an overview level, but that proves to have immense complications when studied closer. The devil really is in the detail. The overview concept is that string theory replaces the messy zoo of fundamental particles with a single entity, a one-dimensional object called a string.

This obviously isn't a real piece of string any more than an electron is a little ball. However, this apparently simple fundamental entity is imagined producing all observed particles, whether matter particles or force carriers, by vibrating in different ways and in alternate configurations, like an open string or a closed loop. After piling layer on layer of abstract mathematics, theoreticians managed to get string theory working—but at a significant cost. Initially there were five major variants of string theory that seemed incompatible, but these were brought together into an overarching theory called M-theory.

String/M-theory faces significant challenges, however. The relatively minor one is that they don't work with the conventional three dimensions of space and one dimension of time. They require nine (string theory) or ten (M-theory) spatial dimensions to work. Dimensions that we clearly don't see. So there needs to be a "fix" like symmetry breaking—in this case by assuming that all the unseen dimensions are curled up so small we don't notice them, though they can still have an influence.

The bigger issue is that string theory is massively open to different possibilities. There are more possible solutions in string theory than there are protons in the universe. There is an inconceivably large set of different outcomes, and the mathematics gives us no way to choose between them. As physicist Martin Bojowald has pointed out, string theory is definitely a theory of everything, because anything and everything can happen in it. And the theory gives no testable predictions to give it any attachment to reality.

British physicist Paul Davies has commented: "[the complexity and lack of predictions] leaves string/M theorists without much of a reality check. Where this enterprise will end is any body's guess. Maybe string/M theorists really have stumbled upon the Holy Grail of science, in which case one day they might be able to tell the rest of us how it works. Or maybe they are all away in Never-Never Land." As some suggest, the dependence on abstract mathematics may have gone too far. Is there a point when mathematically derived theory becomes fantasy?

There is no inherent reason for mathematics to bear any resemblance to the physical world. Those first, stumbling uses of numbers whether counting goats or bags of grain might have had direct real-world analogs, but it soon became clear that interesting math could be done with quantities like negative numbers and their square roots that did not really have an existence in the world around us. To the pure mathematician, engaged in knot theory, for instance, which involves knots unlike any the real world has ever seen, or topology, where there is no difference between a donut and a teacup, this doesn't matter. It is only the challenge of proving relationships and deriving results that is significant. And as long as we don't try to pour tea into a donut we should be fine.

In a way, this freedom is highly empowering. None of our real-world constraints need apply to the pure mathematician. Don't like $2+2=4$? Find it a trifle boring? Then let's make $2+2=5$ and think through what the consequences might be. It might not hold true with oranges, but it is perfectly possible in math. Similarly the limitations of three dimensions of space and a fourth of time have long been arbitrary to mathematicians. It can often be useful to work in "phase space" that has as many dimensions as an object has possible states—potentially trillions upon trillions of dimensions. These dimensions don't exist in the real world, but play a useful role in mathematics. And when mathematicians pushed physics in the direction of string theory, it seemed irrelevant that this involved having nine or ten spatial dimensions, even if technically they appeared not to exist.

Perhaps it's time to bring up Popper. The philosopher of science Karl Popper is somewhat out of fashion with the leading lights of science these days. This is because Popper himself put forward an extreme version of his ideas on the nature of science, saying that science should not use inductive inference. This is where we make predictions based on existing but incomplete observations. So, for example, we infer that light travels at a certain speed because we have always observed it do so. According to Popper, this isn't good enough because it could change tomorrow. Without inductive inference we could hardly do science, and this aspect of Popper's ideas is clearly impractical. But this doesn't discredit a kind of Popper Lite, based on another aspect of his philosophy of science.

This was Popper's claim that a scientific theory had to be capable of being refuted by observation. The argument against this part of Popper's work is usually that we would reject theories too often if they were falsified by any observation to the contrary, so we have to be more sophisticated in its application, in knowing when and where to dismiss a theory. There clearly are limits to the application of Popper's falsification mechanism. The falsification needs to be checked and reproduced before it becomes a serious challenge. But in the end it doesn't stop this being a very powerful assertion that is often necessary to deal with what I'd call invisible dragons—and that also is relevant to a swath of modern physics that has been derived pretty well solely from mathematics.

The invisible dragons argument shows why science needs to have Popper's requirement for refutation in its armory. Imagine that someone said "I have invisible dragons in my garage" and asked a scientist to verify this claim. The scientist clearly can't see the dragons, and can't necessarily find them by feeling around, both because it could be risky, dragons being what they are, and also because the dragons could just fly out of the way. So the scientist might scatter flour on the floor to detect the dragon's footprints. But the dragon owner points out that her dragons are special massless animals. This means that they won't leave footprints. So the scientist sets up infrared detectors. But the dragon owner points out that her dragons are perfectly insulated. So the scientist looks for disturbances in the air caused by the dragon moving—but the dragon owner points out that her dragons move by quantum tunneling and don't displace air at all.

The result of this inability to make an observation is that it is impossible for science to say anything about the invisible dragons. It doesn't mean that they don't exist. They may indeed exist and have all these capabilities that prevent detection. But if it is not possible to disprove their existence through observation, the dragons have to fall outside science. And this would be true even if the dragon owner could produce the most elegant mathematics that seemed to show that there *should be* invisible dragons in her garage.

It might seem odd that Popper required it to be possible to *disprove* a theory. Here, what the owner hoped to do was to prove that there were dragons in her garage. But disproof is the only possible certainty. If paw prints had been found, there could have been another cause. Perhaps the owner faked them with dragon footwear when the scientist wasn't looking. But if dragon theory made a testable prediction and that could be shown to definitely be wrong, then we would know that dragon theory was incorrect. That's unless the theory were changed, incorporating the failed prediction, as often happens with real science, which makes the Popper approach sometimes difficult to apply in practice. This doesn't, however, stop it from being useful.

Science generally isn't capable of producing a proof. It doesn't work in "fact" but in "theory best supported by the evidence." The traditional example of this is the black swan. For hundreds of years in Europe it was assumed that all swans were white, because every swan that was observed was white, which meant

"All swans are white," was a good theory supported by all the available evidence. But this didn't make it a scientific proof. You *can't* prove that all swans are white, just that all the swans that you have ever seen are white. And as soon as someone produced a black swan from Australia it became possible to definitively disprove the theory "all swans are white." (Or at the very least, move to the more sophisticated theory "All European swans are white.")

Similarly, if we look at the big bang theory in cosmology, we can never *prove* that this is an accurate model of the origin of the universe. However, it would be easy to come up with evidence that disproved it. Since the 1950s this has, indeed, happened a number of times, requiring cosmologists to patch up and alter the theory to match the new evidence. (The astrophysicist Fred Hoyle bitterly pointed out to the last that his opposing steady state theory was never given the same opportunity to be patched up to match new data, which he demonstrated admirably could be done.) At the moment, with a few issues, the big bang theory matches the data very well—but it is always capable of being disproved, so it stands up as a good, Popper Lite approved theory.

The same can't be said for theories that are constructed from layer upon layer of mathematics without any strong reference to reality. In Popper Lite terms, string theory is not yet (and may never be) an acceptable scientific theory. This doesn't mean that it is not worth investigating. That could result in discovering predictions that the theory makes that enable it to be falsified. But at the current state of development of string theory, after hundreds of scientists have dedicated decades to it, there is still no way to disprove it. The name really should arguably be the "string hypothesis" at best. A theory should be testable, where a hypothesis is more of a suggestion that as yet does not have the need for rigorous checking.

If you play with language it is possible to set up a scientific theory in such a way that it isn't possible to use the Popper test. If I made the swan example into the theory "Black swans exist," then it can't be disproved. But when I produce a black swan, I have proved the theory. But that reflects the simplicity of this particular theory—this kind of reverse-wording approach is only possible with very simple concepts. In this case, all that I am testing is whether or not a label is valid, and I can do so by direct observation. But when we get to fields like particle physics or cosmology, all the evidence is indirect. I can't prove something exists because it is impossible to get my hands on it and experiment with it. (This indirect nature was true, for instance, with the hunt for the Higgs boson.) It is in these circumstances that the need to be able to disprove a theory becomes essential.

The nagging doubt that there had to be some connection between the mathematics and reality was tucked away in string theory by saying that the extra dimensions were rolled up, too small to see, though this doesn't deal with the vast number of solutions. But had math finally become too far removed from reality? Had science forgotten what it was supposed to do: to be our means of understanding and explaining the workings of the universe?

15 Cargo Cult Science?

As we come toward the end of our exploration of the relationship between mathematics and reality, it is hard to avoid the suggestion that, in physics at least, the tail now wags the dog—it is mathematics that has the upper hand, and the outcome can easily make for an uncomfortable separation between what scientists consider everyday and what makes sense to everyone else.

The physicist Eugene Wigner told a story where two high school friends are discussing their careers. One has become a statistician, and proudly tells his friend about his work. The statistician produces a paper describing the way populations change with type and shows his friend the way that a particular type of distribution, the Gaussian distribution, is used to make predictions about those populations. Inevitably he has to explain the various obscure symbols that crop up in the paper along the way.

The friend is suspicious. He can't see how the statistician could possibly know that the chart he has drawn in his paper, really just a visual representation of an abstract collection of numbers, could somehow predict how a human population, a group of living, thinking individuals, could behave. But he discovers there is worse still hidden among the mathematics. The friend points to one symbol in the paper and asks what it means. "That's pi," says the statistician. "You know what that is. The ratio of the circumference of a circle to its diameter." The friend shakes his head. "Now I know you're messing with me. What has the population got to do with the circumference of a circle?"

Wigner tells a version of this story as an introduction to an attempt to explain what he describes as the "unreasonable effectiveness of mathematics in the natural sciences." This comes up in two ways, because mathematics works where there seems no obvious reason that it should (such as the surprise occurrence of pi in the behavior of people) and secondly, just because the mathematics fits the same pattern, it does not definitely mean that there is any connection between the mathematics and reality—it could be a coincidence, or it could be that there could be multiple matches, and this just happens to be the one we've tried, which works so far, but might stop working tomorrow or when we try to apply it to a different example.

It's useful to go back to the very beginning, to the creation of mathematics. In the opening chapters we saw how numbers were likely to have first come into play as representation of actual physical objects. Numbers that may have originally been simply a match of fingers to goats were probably abstracted first to a match of the same fingers to other things, and then abstracted even further to symbols that stood in for those fingers. But at this stage, what was being dealt with was still something with a clear and direct relationship to physical objects. As math developed we took a step back with negative numbers—just about conceivable to represent how many objects have been removed from a larger collection—and then off into the wilderness with anything from imaginary numbers to aleph-null. Wigner notes:

> Most more advanced mathematical concepts . . . were so devised that they are apt subjects on which the mathematician can demonstrate his ingenuity and sense of formal beauty.

What we see mathematicians doing, and can at the same time be both amazed and baffled by, is pushing the bounds of what is possible, setting limits to keep their math legal within the framework they are employing, and generating a series of concepts that fit together logically even if that whole structure has no application or parallel in the real world. This process is amazing and beautiful, because the mathematicians—mere human beings—are building worlds, managing to achieve so many consistent results (even if this sometimes involves tweaking the rules, as when 1 ceased to be a prime number). The process is also baffling to the outsider because it is not always clear why anyone should put effort into a particular topic, unless it is a vehicle to show off how clever they are.

As with science, though, it is not practically possible to steer mathematics purely in a practical direction. We need to give mathematicians the freedom to experiment as we never know when an obscure piece of mathematical chicanery will result in a practical tool. This often provides a real difficulty for politicians and others outside of science and mathematics who are responsible for funding. It looks like they are paying people to play—and in the case of the pure mathematicians, paying them to play with abstract and irrelevant concepts. But we just don't know when they may come in useful.

Mathematicians and scientists are like hoarders, tucking away all sorts of apparent junk because it might come in handy some day. And when it does, it can have a major impact. Few people outside of mathematics—certainly not Albert Einstein—had any real awareness of what might be possible with non-Euclidian geometry (see page 208) at the start of the twentieth century. Outside of navigating on a curved Earth it seemed purely abstract. But then it was needed to complete the general theory of relativity.

Mathematicians are perfectly entitled to pursue their trade without ever coming close to reality. If we try to justify what they do based on applications, it's a bit like trying to justify the manned exploration of space based on spin-offs—the benefits that society has accidentally gained as a result of the investment of thought and money needed to get people into space. Some of these are genuine, though the impact is usually relatively minor. Some of the suggestions don't follow through logically. I have seen the likes of GPS satellite navigation, weather satellites, and space telescopes all being put forward as a justification for manned space flight, but in reality these could have been done, and mostly have been done, apart from one maintenance mission to Hubble, without any expensive and dangerous human missions.

Finally there are the justifications that depend on spin-offs that are just downright wrong. I've seen it said that it's thanks to NASA that we have the nonstick material PTFE, used everywhere from frying pans to plumbing, Velcro, and personal computers. In fact the first two predate NASA by decades, coming from perfectly normal research and development—they just happen to have been used by NASA. And while it's true that the NASA engineers did have to think about ways to shrink a computer to go in a space capsule, by far the biggest driver for personal computing was not this, but the potential of a mass market, much more significant in making things happen than small-scale specialist applications. Instead, we have to forget the spin-off "benefits" and take manned space travel for what it is: a glorious adventure and a potential survival mechanism for humanity.

Similarly, although we can try to justify what mathematicians do in their explorations of abstract mathematics by looking at the "spin-off" applications of pure math, it isn't why most mathematicians do their research. Such a focus on applications probably wouldn't bring the biggest and most important breakthroughs. Mathematicians do what they do for the challenge and for the fun of the mental achievement. Physicists, though, usually have ties to reality that should prevent them from spending too long on pure flights of fantasy. Although in principle we could envisage a breed of xenophysicists (perhaps they exist, though my spell-checker doesn't think so), who explore and construct the physics of wholly theoretical worlds—and though much actual physics work is done on models of the universe that are so simplified that they bear little resemblance to reality—the role of physics is to predict and to explain the behavior of nature. It should always have the same tie to reality as the goat-counter's fingers do to actual, smelly goats.

A starting point to understanding what physics really does is to get a better feel for the concept of models. There is an old joke, popular among scientists that I retell at every possible opportunity, both because it illustrates perfectly the nature of scientific models, and because it is a great way to tell the difference between people who understand science and those who don't. The ones who don't understand science at best laugh politely. The joke features a geneticist, a dietician, and a physicist, arguing about the way to produce the best racehorse. The geneticist says, "We need to breed horses through many generations, selecting for the right characteristics until we have the perfect runner." In response, the dietician says, "No, it's more a matter of ensuring that the horse has the perfect nutrient balance for muscle growth as it matures." The physicist shakes her head. "Let's assume the racehorse is a sphere."

This kind of simplification in a model works well as a joke, but there's more than an element of truth in it. I remember being irritated by physics problems at school that started by saying, "Assume there is no friction or air resistance." That seemed to be cheating, because there *is* friction and air resistance. You might as well say, "Assume I know the right answer." However it reflects the fact that physicists have to be far more pragmatic than mathematicians, who have perfect control over their numerical universes. Think, for example, of the physics of an object you should be very familiar with—your body. There has to be simplification, rather than dealing with every single atom individually, both because atoms are quantum particles and have a probabilistic nature, and also because there are so many of them. Your body contains somewhere around 10^{27} atoms: 1,000 trillion trillion atoms. No one could ever sensibly make predictions about the behavior of each atom, so we have to take the whole as a lump and model that.

Given the complexity of everything around us, what is surprising is that physics manages to make any predictions at all about how things behave. And yet it does, primarily via simplified mathematical models. In principle these don't have to work. But as we have seen, the universe conspires to help the physicist by, generally speaking, being consistent. If, for instance, values that we regard as fundamental constants like the speed of light were constantly changing in value, it would be impossible to do physics. Leaving aside the disastrous implications for all kinds of physical systems, we couldn't say how anything would behave, because this afternoon it could be totally different from the way it was this morning.

We also expect that things that work on Earth will also work the same elsewhere. That doesn't seem to be reflected in reality. We know, for instance, that things weigh a lot less on the Moon than they do back home. So machines, for instance, will behave differently. But one of Newton's great innovations was the daring assumption that gravity was *universal*. That it worked the same way on the Moon as it does on the Earth. And to date, with a few possible exceptions, this seems a valid assumption. It's not a provable fact—but unless we make assumptions of this kind and see where they take us, there is no point attempting scientific endeavors.

Thankfully we seem to be able to get away with this kind of assumption a lot of the time, and that means that we can be surprisingly successful with our application of mathematical models to reality. The principle of invariance can often be applied—which generally means that a physical "law" that we model will work wherever it is applied in the universe of time and space.

What is less obvious, and causes all kinds of confusion with those who are enthusiastic about, say, homeopathy or who believe that they are sensitive to electromagnetic radiation or see ghosts, is that although science is invariant on who performs the experiment—it doesn't matter if it's a man or woman, young or old, of any ethnicity or race—not all "experiments" are of the same quality. They don't all have the same degree of control. Professional scientists put a huge amount of effort into isolating a phenomenon so that the data emerging from it can't be confused with the impact of something else. This is relatively easy to do in the controlled environment of a laboratory, but is almost impossible to do perfectly outside of a controlled space.

This is why, for instance, in 2014–15 the BICEP2 experiment results, which were first thought to have detected gravitational waves, were later dismissed as an effect of dust in the Milky Way. Any astronomical or cosmological observation has the real potential for unexpected factors to play a part. And the same is true of everyday experience. There's an old saying in science that "the plural of anecdote is not data." Just because we experience something, does not mean whatever interpretation of that experience springs to mind is a good scientific model. If I feel better after taking a homeopathic remedy, I have no way of knowing if that remedy was the cause of my recovery, because I have no controls in place. There could have been many, many causes. Most likely I would have gotten better anyway. Or I just feel that I am better, but have not undergone any actual physical change. So we have to be aware that the assumption of invariance does not mean that an anecdote provides scientific evidence.

As Eugene Wigner points out, "[T]he laws of nature can be used to predict future events only under exceptional circumstances—when all the relevant determinants of the present state of the world are known." Generally speaking this never applies. We don't know *everything* about the circumstances we are dealing with, and when that is the case, we have to make assumptions and simplifications. Our models become more abstracted from reality and are more likely to produce incorrect results.

Even so, in some areas as we have seen, mathematics has proved a surprisingly effective tool, producing results where there is no model based on analogy, but rather it is abstract mathematics that seems to be making a prediction that happens to match reality. Wigner gives the example of the Lamb shift, a small difference in two electron energy levels in an atom that was one of the steps toward the development of quantum electrodynamics. As Wigner puts it: "The quantum theory of the Lamb shift, as conceived by Bethe and established by Schwinger, is a purely mathematical theory and the only direct contribution of experiment was to show the existence of a measurable effect. The agreement with calculation is better than one part in a thousand."

However, the success of such approaches is in no sense an assurance that this kind of modeling will always be possible. It is always an empirical matter. At the moment, for instance, we have two different mathematical structures dealing with quantum theory and relativity, the two great pillars of physics that emerged in the twentieth century. These structures cannot be merged. The assumption made by pretty well all physicists is that there will be a way to do it, either by changing one theory or the other so that it *will* fit with the other's mathematical structure, or by coming up with a new structure that can cope with both. However, it is entirely possible that these will remain the best workable models and that no union, no "theory of everything" can ever be reached.

There doesn't have to be a universal mathematical system that can cope with both. Despite claims to the contrary I would argue that the universe is not inherently mathematical, except in the sense that some (but not most) mathematics is based on an observation of reality. Instead, mathematics is a great tool with which to build models of the universe. Models that will always have limitations. To quote Wigner again: "[F]undamentally, we do not know why our theories work so well. Hence, their accuracy may not prove their truth and consistency." As Wigner so powerfully points out, just because a mathematically based theory gives excellent predictions does not necessarily make it of any value.

That seems bizarre. How can mathematics make an effective prediction, but not be in any sense an accurate representation of reality? Because, as we've already discovered, almost all physics depends on immense degrees of simplification and assumption. We could be missing huge swaths of what actually lies beneath the surface of the black box universe that we are modeling. Let's take a trivial example. Here's a computer program to predict whether the Sun will rise tomorrow morning:

```
IF YEAR < 3000
PRINT "Sun will rise"
ELSE
PRINT "Sun will not rise"
END
```

That is effectively a mathematical model, it's just that most of us are more familiar with the way computer programs go about logic than we are with the symbolic structures of mathematics. I can say that this model will predict what will happen with unerring accuracy until the year 3000, when (I hope) it will go wrong. You might argue that there is no connection between the specific date I have put in and reality. But that, in a sense, is the whole point. We don't know that there is a connection between the constants and formulas we put into scientific equations and reality either. All we can say is that there has been a good match with reality where and when we have looked. But the contents of those equations don't necessarily have any more connection to reality than does my impressively accurate but worthless program.

Throughout this book we have met a series of remarkable mathematicians who have brought new and powerful mathematical concepts into being. Sometimes these were developed as an exercise in pure mathematics. When Girolamo Cardano dreamed up the imaginary number back in the sixteenth century, for instance, he had no idea how it would be used. But imaginary numbers became a powerful tool in the hands of scientists and engineers. Less often, mathematical techniques have been developed with the conscious intent of producing a new practical method, most obviously in Newton's method of fluxions that became calculus.

In establishing just how real mathematics is, whether numbers are truly real, it can be useful to think of mathematicians like the blacksmiths of old. Blacksmiths were the tool builders of the community. They made the essential equipment for the other trades. But they were also artists, finding new and interesting ways to work metal that did not necessarily have a clear application. Similarly mathematicians have acted as tool builders for the sciences, even though they may themselves have been more interested in the abstract world of pure mathematics.

Despite all the successes science has had with math, it is possible for assumptions about the applicability of mathematics to be taken too far. The American physicist Richard Feynman once referred to "cargo cult science," which happened when pseudoscientists followed the form of scientific investigation, but didn't do real science. However, I think there's a better use for the term. The cargo cults of the Melanesian islands supposedly made models of real things like airplanes in the first half of the twentieth century and may have confused them with reality (the history is rather less certain than the myth is on the subject of cargo cults), hoping they would attract real airplanes—and I think that some scientists also confuse *their* models with reality.

The mathematics that was supposed to be a tool to help us understand and explain the physical world has become a free-standing entity, producing results that we then have to scramble to fit to the observation. Despite hundreds of highly intelligent people trying to do this with string theory, for example, it seems entirely possible that they will never succeed. As physics professor Sabine Hossenfelder puts it: "Somewhere along the line many physicists have come to believe that it must be possible to formulate a theory without observational input, based on pure [mathematical] logic and some sense of aesthetics. They must believe their brains have a mystical connection to the universe and pure power of thought will tell them the laws of nature."

Of course, physics is not all there is to science. Other disciplines are still catching up in terms of the use of mathematics. As we have seen (see page 137), all too often those working in the soft sciences, like psychology, struggle with apparently simple, but deceptively counterintuitive tools like statistics and produce erroneous results. There are parts of science where the understanding of math still has to be improved to properly be able to model what is observed. But in physics, perhaps it is time to indulge in a little more stamp collecting and a little less of the ivory-tower mathematics.

This is not a Luddite cry. Mathematics that is beyond the abilities of 99 percent of us is needed to understand the theory behind the quantum technology that is deployed in everyday devices like smartphones. But equally, we perhaps need to be looking for different, better explanatory models that derive more from experiment and less from mathematical theory. Because, in the end, science is not purely about modeling reality at a numerical level. It is also about constructing models that reflect the observations we make of the world around us to help all of us understand it. And there is never a single way of doing this.

Statisticians have long warned that correlation is not causality. Just because two sets of data move together over time does not mean that one is the cause of the other—nor does it mean that this correlation will continue into the future (which requires a causal link to be sure of the continuance). A classic example of correlation without causality was that after the Second World War, imports of bananas into the UK were well correlated with pregnancies. As the number of bananas went up, so did the pregnancies (and vice versa). No one sensibly believes that the pregnancies were caused by the bananas (in practice, both were probably linked to a third factor), but generally speaking the correlations we encounter seem far less ridiculous, so more likely to be accepted as a causal link at face value.

There are plenty of readily available weird correlations that certainly aren't causal. There is even a website that produces them automatically, where you can learn that the number of suicides by hanging was very strongly correlated with U.S. spending on science, space, and technology over a ten-year period, or that the per capita consumption of margarine strongly followed the divorce rate in the U.S. state of Maine over a similar period of time. And this strong mathematical linkage is obviously meaningless. Yet we rarely raise an eyebrow when a news anchor tells us that the stock market has gone down as a result of new government policy—even though, once again, the causal relationship is only assumed.

One of the reasons that scientists are so enthusiastic to perform tests in laboratory conditions is that it enables them to keep far more of the possible variables under control than would be possible "in the wild." When researching my book on powers of the mind, *Extra Sensory,* it became very obvious that there was a huge difference between claimed abilities in properly controlled conditions and in a performance environment. In the lab causality is relatively easy to establish, but on the stage it is much easier to produce correlation where misdirection hid the true cause of the phenomenon.

This difficulty of pinning down causality is why it can be so tough to get a definitive story on the impact of different diets on human health. Scientists might notice, for instance, that people with a diet rich in tomatoes have a lower chance of heart disease. But we can't assume that all we need to do is increase our tomato consumption to improve our health. Because in the complex world we inhabit day to day, as opposed to a highly controlled experiment, we will find that people who have a diet rich in tomatoes have a whole host of other differences from, say, someone who eats a lot of junk food—and it may not be the tomatoes that are making the difference.

If we could put thousands of people in a lab and control what they eat for months, then we could have rigorous scientific analysis, but as it is, most diet studies have to take a whole mess of differences together (and often also have to rely on notoriously inaccurate self-reporting) and so struggle to achieve scientific accuracy.

Things aren't quite so bad for a physicist working in a lab, but the same kind of issues dog the kind of science we can't control so easily—like cosmology. Even in the lab there can be plenty of circumstances where the situation being studied is complex, or observations are highly indirect—such as the complex and messy results from the detectors of the Large Hadron Collider. In such circumstances, it can be tempting to fit a mathematical approach because it "looks right" and is aesthetically appealing, rather than because there is any observation-driven reason for applying that mathematics in the first place. The result can be to produce a mathematical model that churns out numbers that match experimental results, but where the model has no linkage to reality. The overreliance on math is something that a number of contemporary physicists have highlighted.

There have been plenty of instances in the history of science when this kind of mismatching occurred. For example, the system of epicycles used in the astronomy of Ptolemy was based on fitting a mathematical model to observation and fine-tuning it to what was observed. The approach would remain in use for more than 1,300 years and did indeed fit observation fairly well, because of that fine-tuning. But there was no good scientific reason for choosing the model of circles rotating in circles that had evolved to describe the way that planets moved. The model was doomed because it was based on an incorrect *physical* assumption—that the Earth was at the center of the universe. After that, it didn't matter how clever the math was at fitting observation. Mathematics alone wasn't enough and it never can be.

A leading physicist who is uncomfortable with the way that mathematics seems to be in the driver's seat, pushing experiment into second place, is Neil Turok, director of the prestigious Perimeter Institute, who recently commented:

> We've been given these incredible clues from nature and we're failing to make sense of them. In fact we're doing the opposite: theory is becoming ever more complex and contrived. We throw in more fields, more dimensions, more symmetry—we're

throwing the kitchen sink at the problem and yet failing to explain the most basic facts.

In essence, Turok is complaining about the science being driven by the math. One way of looking at the difference between mathematics and science is that mathematics is fundamentally about truth and facts. You can't argue with a fact within a mathematical system. In our traditional arithmetic 2 + 2 will always be equal to 4. It is a fact. It follows from the rules. There can never be evidence to dispute this fact, because it emerges from the nature of the mathematics system used. As the scientist and science-fiction writer Isaac Asimov said: "As time goes on, nearly every field of human endeavor is marked by changes which can be considered as correction and/or extension . . . Now we can see what makes mathematics unique. Only in mathematics is there no significant correction—only extension."

It's important to be careful with the words. It is perfectly possible for 2 + 2 to be equal to something else, if we use a different number base (to base 3, for instance, 2 + 2 equals 11) or if we use the arithmetic we met in chapter 1, where instead of numbers constantly growing with addition, they reach a maximum value, then start again, like the numbers on a clockface. And as far as a mathematician is concerned, neither conventional nor clock arithmetic is more "real," even though one applies to all conventional physical objects while the other is only relevant to cyclical events in the real world. But within the specific system we use for conventional arithmetic, it is not possible to move away from the facts that the system is based on.

Science, however, is not like that. When scientists speaks about the big bang occurring 13.7 billion years ago, starting the expansion of the universe, they are not describing the same kind of fact as is presented when we write $2+2=4$. The big bang is our best theory, based on current data. The particular theory that has that "best" accolade at the time of writing has been revised at least three times when data proved an earlier version was wrong. And it may well be that the whole thing will get thrown out and replaced with something better at a later date.

Science is always provisional. It is not about finding an absolute truth. It is not about facts at its heart. This doesn't mean that facts don't play a huge part in science. In the "stamp collecting" aspects of science that Rutherford was so dismissive about there is a lot of fact collection, in part because a lot of what is being done is assigning labels—constructed facts. And there are often observational facts, whether it's the fact that there is one and only one keyboard under my fingers as I type or that my computer uses energy. It is when science provides explanatory theories that we need to be aware that facts have left the room. If I say light is a wave or a stream of particles or a disturbance in a quantum field, I am not describing light, I am describing a mathematical theory or an analogy. It's one thing to say that light or atoms exist. It's quite another to have a theory that describes what they are or how they work, because they operate on a scale and in an environment that is entirely different to the macro world we observe.

Scientists frequently fail to mention the indirect relationship between science and the truth, or even forget it. This is probably because there is a danger of making it sound like all ideas, all theories, have the same weight. As a science writer I am constantly being sent theories by people who believe that they have shown that Einstein was wrong about relativity. And other people believe in magic that totally disregards science, whether they think that energy can be produced from nowhere or that homeopathy works. But science does a much better job than giving every possibility equal weight. People are very welcome to challenge Einstein—but for the moment, relativity delivers extremely well, and to shift to some new theory would require new and convincing evidence of a theory that better matched experiments and observations of the universe. To dismiss science altogether and move to magic would require even better data.

I would suggest that this shows us why those who believe that science *is* mathematics, or that the universe is mathematical in nature are wrong, even if some of them are immensely clever and intellectually superior to the rest of us. Because the two disciplines are inherently different in this way. One (math) is a collection of facts, which we are able to establish because we fix the rules, and the other (science) is a collection of models and theories, which we can test against data, but can never call the actual truth.

An important factor to consider in establishing just how much weight we should give to mathematics that is divorced from the world is that while it is possible for mathematicians to weave their magic in splendid isolation from the real world, and to truly base everything on mathematics, this is not really how the application of math to science works. As mathematician Richard Hamming, who we met in chapter 2, comments:

> We select the kind of mathematics to use. Mathematics does not always work. When we found that scalars did not work for forces, we invented a new mathematics, vectors. And going further we have invented tensors . . . we select the mathematics to fit the situation, and it is simply not true that the same mathematics works every place.

Science has done remarkable things and will do even more. Mathematics has proved a remarkable tool as a method for reliably modeling aspects of the universe, and will continue to be a wonderful tool. Using math in an appropriately scientific manner remains by far the best way of understanding the fundamentals of physics and the universe—this is not a call for a free-for-all where anything goes. It is equally important that the scientific community does not confuse models with reality, but always remembers that mathematics has many worlds, which may, like a mirror image, sometimes succeed in reflecting reality. This does not make the two the same thing. There is no mirror mathematical world for Alice to go through the looking glass and visit.

Some numbers and mathematical processes are, without doubt real, or at least have a one-to-one correspondence with real objects and actions. The natural numbers, the nonnegative integers were definitely originally derived from physical objects, and undergo exactly the same arithmetical processes as real objects do. It took a while for negative integers to be so grounded, but they can be seen in operation in the arithmetic of electrical charges. As soon as we move further, the gap between mathematics and reality becomes clearer. Both fractions and geometry, for example, though having very clear parallels in the real world, are refinements that can never truly match reality's messiness. Like Plato's cave, we inhabit a world where it isn't possible to truly have geometric shapes with zero width lines, nor can we make a true, perfect division of cake into perfect fractional pieces because of the grainy, atom-based nature of reality.

Admittedly we can divide something like money in exact fractions . . . sometimes. So 25 cents is exactly and precisely a quarter of a dollar. But that is because money is quantized by definition. It comes in ready-made divisions that allow us to apply those fractions. However, such an approach has its own mathematical cost. We can indeed produce a perfect quarter of a dollar, but there is no mechanism to produce a third of a dollar. It simply does not exist in reality, even though it is easy enough to represent mathematically.

As we have seen in our journey in this book, from the mathematician's viewpoint, fractions and geometry are just the beginning of vast landscapes of awesome mathematical power and splendor. They are landscapes where mathematicians can spend their entire lifetimes exploring without ever coming across anything that could be considered to be real. Sometimes, though, the mathematical structures and mechanisms do parallel the real world. Such numbers and procedures may not actually be real, but they can still help answer our questions.

Despite its ability for abstraction, we need to keep our practical mathematics grounded in the physical so that the language of science can speak to us all. To come back to the question in the title of the book—numbers, I would suggest, *are* real at their most basic, but most of mathematics isn't. It's a fantasy world that sometimes mirrors and parallels our own, and as such can help provide us with tools to help understand reality. But it needs to be kept in its place. And as long as we (and scientists) remember this, we can't go far wrong.

Notes

1. Counting Sheep

4—The article on poor attitude to math in the UK is Wendy Jones, "Bad Attitudes to Maths Makes Children Switch Off," *Guardian,* March 7, 2012, accessed July 12, 2015, www.theguardian.com/teacher-network/teacher-blog/2012/mar/07/world-maths-day-adult-innumeracy.

5—St. Augustine's damnation of mathematicians is from St. Augustine, *Genesi ad Litteram* (*The Literal meaning of Genesis*), trans. Roland J. Teske (Washington: CUA Press, 2010), p. 76.

6—The British Court of Appeal discussion of the meaning of the number one is reported in Steve Connor, "What Exactly Does 'One' Mean? Court of Appeal Passes Judgement on Thorny mathematical Issue," *Independent,* accessed June 30, 2015, www.independent.co.uk/news/science/what-exactly-does-one-mean-court-of-appeal-passes-judgement-on-thorny-mathematical-issue-10350568.html.

9—Steven Weinberg's comparison of the International Space Station and the Superconducting Super Collider was made in a telephone interview with the author in February 2013.
10—Max Tegmark's assertion that the universe is made of mathematics is from Max Tegmark, *Our Mathematical Universe* (London: Penguin Books, 2015), pp. 243–318.

2. Counting Goats

15—The Ishango bone is described in Amir Aczel, *Finding Zero* (New York: Palgrave, 2015), pp. 20–21.
17—Details of the variants of Uruk number systems from Leonard Mlodinow, *The Upright Thinkers* (London: Allen Lane, 2015), p. 49.
18—Richard Hamming's amazement at the abstraction of integers is described in Richard Hamming, "The Unreasonable Effectiveness of Mathematics," *American Mathematical Monthly* 87, no. 2 (February 1980), pp. 81–90.
22—Information on Ancient Greek numbering from Amir Aczel, *Finding Zero* (New York: Palgrave, 2015), p. 14.

3. All Is Number

30—Theories for the reason the Pythagoreans avoided beans are discussed in Charles Seife, *Zero: The Biography of a Dangerous Idea* (London: Souvenir Press, 2000), p. 25.
30—Details of the meanings ascribed to numbers by the Pythagoreans from Brian Clegg, *A Brief History of Infinity* (London: Constable & Robinson), pp. 34–35.

31—The prediction of a tenth heavenly body by Philolaus is described in Steven Weinberg, *To Explain the World* (London: Allen Lane, 2015), p. 78.
37—Details of Zeno's paradoxes from Brian Clegg, *A Brief History of Infinity* (London: Constable & Robinson, 2003), pp. 10–17.
38—The Greek consideration of numbers as collections of objects and fractions as parts is described in David Fowler, *The Mathematics of Plato's Academy* (Oxford: Oxford University Press, 1999).

4. Elegant Perfection

47—Information on Euclid's work from *Euclid's Elements: All Thirteen Books Complete in One Volume*, trans. Thomas Heath (Santa Fe, NM: Green Lion Press, 2002).
52—Details of the Ancient Greeks who spent their time attempting to square the circle from Brian Clegg, *A Brief History of Infinity* (London: Constable & Robinson, 2003), p. 68.

5. Counting Sand

56—The description by Plutarch of Archimedes considering mechanical devices to be geometry at play is quoted in Thomas Heath, *The Works of Archimedes* (New York: Dover, 2002), p. xvi.
57—The suggestion that Archimedes was related to the royal family of Syracuse is from Thomas Heath, *The Works of Archimedes* (New York: Dover, 2002), p. xv.
57—The quote from *The Sand-reckoner* on sand location is from Thomas Heath, *The Works of Archimedes* (New York: Dover, 2002), p. 221.

58—Archimedes quote from Aristarchus on the size of the sphere of the fixed stars is taken from Thomas Heath, *The Works of Archimedes* (New York: Dover, 2002), p. 222.

6. The Emergence of Nothing

65—The views of Brahmagupta and Bhāskara on zero divided by zero are from Brian Clegg, *A Brief History of Infinity* (London: Constable & Robinson, 2003), p. 111.
66—The introduction of the BC/AD dating system is discussed in Bonnie Blackburn and Leofranc Holford-Strevens, *The Oxford Companion to the Year* (Oxford: Oxford University Press, 2003), pp. 767–82.
68—The Old Babylonian tablet with a value for the square root of 2 is described in Carl B. Boyer and Uta C. Merzbach, *A History of Mathematics* (New York: John Wiley & Sons, 1991), p. 27.
71—The mention of Hindu numerals by a Syrian bishop in 662 is noted in David Eugene Smith, *History of Mathematics*, vol. 1 (New York: Dover, 1958), p. 167.
72—The translation of the Cambodian inscription with a zero dating to AD 683 is in Amir Aczel, *Finding Zero* (New York: Palgrave, 2015), pp. 94–97.
73—The story of Amir Aczel's successful attempt to find the oldest known zero is in Amir Aczel, *Finding Zero* (New York: Palgrave, 2015), pp. 98–178.
73—The original Indian use of the same symbol and concept for a placeholder zero and for an unknown quantity is described in Robert Kaplan, *The Nothing That Is* (London: Penguin, 2000), pp. 57–59.
74—The different interpretations of zero from classical Indian mathematicians are taken from Robert Kaplan, *The Nothing That Is* (London: Penguin, 2000), pp. 72–73.

75—John Donne's railing against "nothing" is from John Donne, *The Works*, vol. 6 (London: Parker, 1839), p. 155.
75—The resistance to use of numerals rather than words in Florence and Belgium is described in Robert Kaplan, *The Nothing That Is* (London: Penguin, 2000), p. 102.
77—The Ancient Greek style word form of the equation A+B=C+D and their approach to numerals is described in Reviel Netz, *The Shaping of Deduction in Greek Mathematics* (Cambridge: Cambridge University Press, 1999), p. 98.
77—The modern rendering of the approach taken by Diophantus to an algebraic equation is from Carl B. Boyer and Uta C. Merzbach, *A History of Mathematics* (New York: John Wiley & Sons, 1991), p. 181.
78—The suggestion that algebra may have been developed to help with complex inheritance rules is in Carl B. Boyer and Uta C. Merzbach, *A History of Mathematics* (New York: John Wiley & Sons, 1991), p. 233.

7. He Who Is Ignorant

79—The quote from Roger Bacon on the importance of mathematics is from Roger Bacon, *Opus Majus*, trans. Robert Belle Burke (Kila, MT: Kessinger Publishing, 1998), p. 116.
88—Information on Muybridge's Zoopraxographical Hall at the World's Columbian Exposition from Brian Clegg, *The Man Who Stopped Time* (Washington, DC: Joseph Henry Press, 2007), pp. 225–26.
88—The book that temporarily ruined Muybridge's reputation as the grandfather of moving pictures was *A Million and One Nights* by Terry Ramsaye, discussed in Brian Clegg, *The Man Who Stopped Time* (Washington, DC: Joseph Henry Press, 2007), pp. 245–47.

89—Roger Bacon's claim that squaring the circle was "clearly understood" comes from Roger Bacon, *Opus Majus,* trans. Robert Belle Burke (Kila, MT: Kessinger Publishing, 1998), p. 15.

90—The quote on Bacon's support of math from John Wallis is from Joseph Frederick Scott, *The Mathematical Work of John Wallis* (New York: Chelsea Publishing Company, 1981), p. 142.

90—Roger Bacon's extended eulogy to the importance of mathematics is from Roger Bacon, *Opus Majus,* trans. Robert Belle Burke (Kila, MT: Kessinger Publishing, 1998), pp. 116–17.

92—Information on Bradwardine and Oresme from Carl B. Boyer and Uta C. Merzbach, *A History of Mathematics* (New York: John Wiley & Sons, 1991), pp. 262–66.

8: All in the Imagination

98—Information on Cardano and Tartaglia from Carl B. Boyer and Uta C. Merzbach, *A History of Mathematics* (New York: John Wiley & Sons, 1991), pp. 282–89.

98—Cardano's remark that an imaginary number was "as subtle as it is useless" is quoted in Roger Cooke, *The History of Mathematics: A Brief Course* (London: Wiley, 1997), p. 310.

9. The Amazing Mechanical Mathematical Universe

102—The Marquis of Laplace wrote about an intelligence that could perfectly predict the future in Pierre Simon Laplace, *A Philosophical Essay on Probabilities,* trans. F. W. Trusscott and F. L. Emory (New York: Cosimo Inc., 2007), p. 4.

113—The assertion that Descartes primarily intended his analytical geometry as a way of constructing geometric forms, not of deducing algebraic forms is from Carl B. Boyer and Uta C. Merzbach, *A History of Mathematics* (New York: John Wiley & Sons, 1991), p. 337.

105—Details of the contents of Newton's library are from John Harrison, *The Library of Isaac Newton* (Cambridge: Cambridge University Press, 2008), p. 59.

110—Newton's letter to Leibniz with his coded claim is from Isaac Newton, *The Correspondence of Isaac Newton,* vol. 2, ed. H. Turnbull (Cambridge: Cambridge University Press, 1959), p. 134.

112—The quote from Berkeley on the errors in calculus is taken from Carl B. Boyer and Uta C. Merzbach, *A History of Mathematics* (New York: John Wiley & Sons, 1991), p. 430.

10. The Mystery of "Maybe"

132—Maxwell's development of a distribution for the speed of gas molecules is described in Rhodri Evans and Brian Clegg, *Ten Physicists Who Transformed Our Understanding of Reality* (London: Robinson, 2015), p. 93.

135—The study showing that a double SIDS case would be expected in England every eighteen months was Stephen Watkins, "Conviction by Mathematical Error?," *BMJ,* 320, no. 7226 (2000), pp. 2–3.

136—Information on Isaac Asimov's use of mathematical prediction in the Foundation series from Brian Clegg, *Ten Billion Tomorrows* (New York: St. Martin's Press, 2015), p. 272.

138—The details of Linzmeyer's fifteen-card correct run and the quote from Rhine on the subject are from Joseph Banks Rhine, *Extra-Sensory Perception* (Hong Kong: Forgotten Books, 2008), p. 86.

11. Maxwell's Mathematical Hammer

146—Biographical details on Maxwell and details of his electromagnetic theory from Rhodri Evans and Brian Clegg, *Ten Physicists Who Transformed Our Understanding of Reality* (London: Constable, 2015), pp. 252–71.

156—Frank Wilczek's assertion that the collection of fields that fill the universe form an ether is made in Frank Wilczek, *The Lightness of Being* (Philadelphia: Basic Books, 2008).

157—Details of retarded and advanced waves from Brian Clegg, *How to Build a Time Machine* (New York: St. Martin's Press, 2011), pp. 152–55.

12. Infinity and Beyond

161—Ancient Greek ideas on infinity are from Brian Clegg, *A Brief History of Infinity* (London: Constable & Robinson, 2003), pp. 29–32.

163—Gauss's dismissal of the reality of infinity was made in a letter to the Danish astronomer Heinrich Christian Shumacher, dated July 12, 1831.

163—The details of Galileo's work on infinity and his attempts to get the book containing it published are from Brian Clegg, *A Brief History of Infinity* (London: Constable & Robinson, 2003), pp. 80–92.

164—Quotes from Galileo's dedication are from the translation by Henry Crew and Alfonso de Salvio of Galileo Galilei, *Dialogues Concerning Two New Sciences* (New York: Dover, 1954), pp. xvii–xviii.

166—Sagredo's assertion about ants carrying a ship is from Galileo Galilei, *Dialogues Concerning Two New Sciences* (New York: Dover, 1954), p. 20.

172—The suggestion that ordinal numbers may predate cardinal numbers is from Carl B. Boyer and Uta C. Merzbach, *A History of Mathematics* (New York: John Wiley & Sons, 1991), p. 5.
174—The idea that basing the natural numbers on set theory abstracts it from reference to the real world is from Roger Penrose, *The Road to Reality* (London: Vintage, 2005), p. 65.
184—Gödel is quoted on the problems of the ZFC axioms in Natalie Wolchover, "To Settle Infinity Dispute, a New Law of Logic," *Quanta Magazine,* November 26, 2013, accessed June 30, 2015, www.quanta magazine.org/20131126-to-settle-infinity-question-a-new-law-of -logic/.
192—The quote from Max Tegmark on infinity is from Max Tegmark, "Infinity Is a Beautiful Concept—And It's Ruining Physics," *Crux,* February 20, 2015, accessed June 30, 2015, blogs.discovermagazine .com/crux/2015/02/20/infinity-ruining-physics/#.VZKZy2DL_gD.

13. Twentieth-century Mathematical Mysteries

202—Einstein's description of having his 'happiest thought' while sitting in the Bern patent office is from his 1922 Kyoto Lecture and referenced in W. F. Bynum & Roy Porter (Eds.) *Oxford Dictionary of Scientific Quotations* (Oxford: Oxford University Press, 2005), p. 198.
207—Proposition 32 of Euclid's Elements is taken from *Euclid's Elements: All Thirteen Books Complete in One Volume,* trans. Thomas Heath (Santa Fe, NM: Green Lion Press, 2002), p. 25.

14. Symmetry Games

225—Albrecht and Magueijo's paper on the possibility of a changing speed of light removing the need for cosmic inflation is Andreas Albrecht and João Magueijo, "Time Varying Speed of Light as a Solution to Cosmological Puzzles," *Physical Review D,* 59, no. 4 (1999), 43516, journals.aps.org/prd/pdf/10.1103/PhysRevD.59.043516.

226—Details of the Oklo natural nuclear reactors and their implications for symmetry of electrical charge through time from Leon Lederman and Christopher Hill, *Symmetry and the Beautiful Universe* (New York: Prometheus Books, 2004), pp. 40–43.

233—Leon Lederman points out the dangers of assuming perfect symmetries in Leon Lederman and Christopher Hill, *Symmetry and the Beautiful Universe* (New York: Prometheus Books, 2004), p. 19.

237—Martin Bojowald's quip that string theory is a theory of everything because everything and anything can happen is from Martin Bojowald, *Once Before Time: A Whole Story of the Universe* (New York: Alfred A. Knopf, 2010), p. 83.

237—The comment that string and M-theorists could be "away in Never-Never Land" is from Paul Davies, *The Goldilocks Enigma* (London: Penguin Books, 2007), p. 47.

238—The concerns about Popper's philosophy of science are taken from Tim Lewens, *The Meaning of Science* (London: Pelican Books, 2015), pp. 22–44.

15. Cargo Cult Science?

243—The story of the statistician whose friend doubts the validity of pi, and other thoughts on the nature of mathematics by Wigner are from Eugene Wigner, "The unreasonable effectiveness of mathematics in the natural sciences. Richard courant lecture in mathematical sciences delivered at New York University," *Communications on Pure and Applied Mathematics* 13, no. 1 (May 11, 1959), pp. 1–14.

254—Sabine Hossenfelder's dismissal of deriving physical theory from aesthetics is from Sabine Hossenfelder (2015). "Does the Scientific Method need revision," accessed June 17, 2015, backreaction.blogspot .co.uk/2015/01/does-scientific-method-need-revision.html.

255—The apparent causal link between import of bananas and pregnancies is discussed in Brian Clegg, *Dice World* (London: Icon, 2013), p. 33.

255—The website providing apparent correlations between different sets of data with no causal link is Tylervigen.com, "15 Insane Things That Correlate With Each Other," accessed June 22, 2015, www .tylervigen.com/spurious-correlations.

256—The book giving details of scientific investigations of psi phenomena is Brian Clegg, *Extra Sensory* (New York: St. Martin's Press, 2013).

256—Details of how dietary science advice has problems in differentiating between correlation and causality are found in Brian Clegg, *Science for Life* (London: Icon Books, 2015), pp. 1–118.

257—Neil Turok's argument that physics is becoming too driven by mathematical theory is from M. Brooks, "Battle for the Universe," *New Scientist* (July, 2015), p. 6.

258—Isaac Asimov's comments on the unique nature of mathematics are from the foreword of Carl B. Boyer and Uta C. Merzbach, *A History of Mathematics* (New York: John Wiley & Sons, 1991), p. vii.

260—Richard Hamming's observation that we select the mathematics to fit the situation is from Richard Hamming, "The Unreasonable Effectiveness of Mathematics," *American Mathematical Monthly* 87, no. 2 (February 1980), pp. 81–90.

Index